Stem Cell Biology and Regenerative Medicine

Series Editor:

Kursad Turksen
Ottawa Hospital Research Institute, Ottawa, ON, Canada

Our understanding of stem cells has grown rapidly over the last decade. While the apparently tremendous therapeutic potential of stem cells has not yet been realized, their routine use in regeneration and restoration of tissue and organ function is greatly anticipated. To this end, many investigators continue to push the boundaries in areas such as the reprogramming, the stem cell niche, nanotechnology, biomimetics and 3D bioprinting, to name just a few. The objective of the volumes in the Stem Cell Biology and Regenerative Medicine series is to capture and consolidate these developments in a timely way. Each volume is thought-provoking in identifying problems, offering solutions, and providing ideas to excite further innovation in the stem cell and regenerative medicine fields.

More information about this series at http://www.springer.com/series/7896

Babak Arjmand
Editor

Genomics, Proteomics, and Metabolomics

Stem Cells Monitoring in Regenerative Medicine

Humana Press

Editor
Babak Arjmand
Cell Therapy and Regenerative Medicine Research Center
Endocrinology and Metabolism
Molecular-Cellular Sciences Institute
Tehran University of Medical Sciences
Tehran, Iran

Metabolomics and Genomics Research Center
Endocrinology and Metabolism
Molecular-Cellular Sciences Institute
Tehran University of Medical Sciences
Tehran, Iran

ISSN 2196-8985 ISSN 2196-8993 (electronic)
Stem Cell Biology and Regenerative Medicine
ISBN 978-3-030-27726-0 ISBN 978-3-030-27727-7 (eBook)
https://doi.org/10.1007/978-3-030-27727-7

This Humana imprint is published by the registered company Springer Nature Switzerland AG
The registered company address is: Gewerbestrasse 11, 6330 Cham, Switzerland

I would like to dedicate this book to the biggest breakthroughs in my life my ***parents****, my loving wife* ***Parisa****, my daughter* ***Rasta****, and my son* ***Arvid****.*

Preface

In the past decade, the science of integrating huge data from genes, RNA, proteins, and metabolites have introduced multiple layers of omics technologies and their potential opportunities in biomedicine. Under the umbrella systems biology, scientists have tried to empower *stem cell therapy and regenerative medicine using omics*. On the other hand, omics can improve basic studies adding some molecular views. A large body of evidence demonstrates that systems biology can provide large-scale and high-throughput approaches to characterize complex biological, physiological, and pathological data of a wide range of molecular pathways. Therefore, exploring the function of individual genes, proteins, and metabolites involved in specific biological processes can be provided by omics platforms. In this regard, the combination of omics strategies with stem cell biology and regenerative medicine boosts the knowledge of molecular mechanisms underlying a variety of incurable diseases and treatments that are suggested based on cell therapy and regenerative medicine. They also can be used to depict cellular fate and pathological progression which allow the most efficient disease diagnosis, predictive assessment of disease relapse, and patient-specific cell therapies.

It is my pleasure having the fingerprints of some prominent scientists in this volume, which could be valuable to both basic and clinical investigators who are interested in the field of molecular and cellular biology, specifically genomics, proteomics, metabolomics, and stem cell and regenerative medicine. I would like to acknowledge *Dr. Kursad Turksen,* Series Editor of the *Stem Cell Biology and Regenerative Medicine*, for his kind advices and support. I also thank *Merry Stuber*, Editor of Cell Biology and Biomedical Engineering, and *Maria David*, Project Coordinator at Springer Nature, for their continuous attempt to get the volume to the print stage.

Tehran, Iran Babak Arjmand

Contents

Contributors

Mina Abedi Metabolomics and Genomics Research Center, Endocrinology and Metabolism Molecular-Cellular Sciences Institute, Tehran University of Medical Sciences, Tehran, Iran

Cell Therapy and Regenerative Medicine Research Center, Endocrinology and Metabolism Molecular-Cellular Sciences Institute, Tehran University of Medical Sciences, Tehran, Iran

Leila Afshar Medical Ethics Department, Shahid Beheshti University of Medical Sciences, Tehran, Iran

Hamid Reza Aghayan Cell Therapy and Regenerative Medicine Research Center, Endocrinology and Metabolism Molecular-Cellular Sciences Institute, Tehran University of Medical Sciences, Tehran, Iran

Sepideh Alavi-Moghadam Cell Therapy and Regenerative Medicine Research Center, Endocrinology and Metabolism Molecular-Cellular Sciences Institute, Tehran University of Medical Sciences, Tehran, Iran

Maryam Arabi Metabolomics and Genomics Research Center, Endocrinology and Metabolism Molecular-Cellular Sciences Institute, Tehran University of Medical Sciences, Tehran, Iran

Cell Therapy and Regenerative Medicine Research Center, Endocrinology and Metabolism Molecular-Cellular Sciences Institute, Tehran University of Medical Sciences, Tehran, Iran

Babak Arjmand Cell Therapy and Regenerative Medicine Research Center, Endocrinology and Metabolism Molecular-Cellular Sciences Institute, Tehran University of Medical Sciences, Tehran, Iran

Metabolomics and Genomics Research Center, Endocrinology and Metabolism Molecular-Cellular Sciences Institute, Tehran University of Medical Sciences, Tehran, Iran

Nikoo Bana Cell Therapy and Regenerative Medicine Research Center, Endocrinology and Metabolism Molecular-Cellular Sciences Institute, Tehran University of Medical Sciences, Tehran, Iran

Khadijeh Falahzadeh Department of Medical Biotechnology, School of Advanced Technologies in Medicine, Tehran University of Medical sciences, Tehran, Iran

Metabolomics and Genomics Research Center, Endocrinology and Metabolism Molecular-Cellular Sciences Institute, Tehran University of Medical Sciences, Tehran, Iran

Najmeh Foroughi Heravani Cell Therapy and Regenerative Medicine Research Center, Endocrinology and Metabolism Molecular-Cellular Sciences Institute, Tehran University of Medical Sciences, Tehran, Iran

Kambiz Gilany Department of Biomedical Sciences, University of Antwerp, Antwerp, Belgium

Reproductive Biotechnology Research Center, Avicenna Research Institute, ACECR, Tehran, Iran

Integrative Oncology Department, Breast Cancer Research Center, Motamed Cancer Institute, ACECR, Tehran, Iran

Parisa Goodarzi Brain and Spinal Cord Injury Research Center, Neuroscience Institute, Tehran University of Medical Sciences, Tehran, Iran

Mahdieh Hadavandkhani Cell Therapy and Regenerative Medicine Research Center, Endocrinology and Metabolism Molecular-Cellular Sciences Institute, Tehran University of Medical Sciences, Tehran, Iran

Mehrdad Hashemi Department of Genetics, Tehran Medical Sciences, Islamic Azad University, Tehran, Iran

Saeed Heidari-Keshel Department of Tissue Engineering and Applied Cell Sciences, School of Advanced Technologies in Medicine, Shahid Beheshti University of Medical Sciences, Tehran, Iran

Masumeh Jalalvand Department of Medical Biotechnology, School of Advanced Technologies in Medicine, Tehran University of Medical sciences, Tehran, Iran

Erdal Karaöz Center for Stem Cell and Tissue Engineering Research & Practice, İstinye University, İstanbul, Turkey

Medical Faculty, Histology and Embryology Department, İstinye University, İstanbul, Turkey

Center for Regenerative Medicine and Stem Cell Research and Manufacturing, Liv Hospital, İstanbul, Turkey

Navaz Karimian Pour Leslie Dan Faculty of Pharmacy, University of Toronto, Toronto, ON, Canada

Fatemeh Khatami Chronic Diseases Research Center, Endocrinology and Metabolism Population Sciences Institute, Tehran University of Medical Sciences, Tehran, Iran

Arash Khojasteh Department of Tissue Engineering and Applied Cell Sciences, School of Advanced Technologies in Medicine, Shahid Beheshti University of Medical Sciences, Tehran, Iran

Bagher Larijani Endocrinology and Metabolism Research Center, Endocrinology and Metabolism Clinical Sciences Institute, Tehran University of Medical Sciences, Tehran, Iran

Ahmad Mani-Varnosfaderani Chemometrics Laboratory, Department of Chemistry, Tarbiat Modares University, Tehran, Iran

Mohammad Javad Masroor Chemometrics Laboratory, Department of Chemistry, Tarbiat Modares University, Tehran, Iran

Arash Minai-Tehrani Nanobiotechnology Research Center, Avicenna Research Institute, ACECR, Tehran, Iran

Solmaz M. Nejad Metabolomics and Genomics Research Center, Endocrinology and Metabolism Molecular-Cellular Sciences Institute, Tehran University of Medical Sciences, Tehran, Iran

Maryamossadat Mousavi Metabolomics and Genomics Research Center, Endocrinology and Metabolism Molecular-Cellular Sciences Institute, Tehran University of Medical Sciences, Tehran, Iran

Babak Negahdari Department of Medical Biotechnology, School of Advanced Technologies in Medicine, Tehran University of Medical sciences, Tehran, Iran

Moloud Payab Obesity and Eating Habits Research Center, Endocrinology and Metabolism Molecular-Cellular Sciences Institute, Tehran University of Medical Sciences, Tehran, Iran

Fakher Rahim Health Research Institute, Thalassemia and Hemoglobinopathies Research Center, Ahvaz Jundishapur University of Medical Sciences, Ahvaz, Iran

Azam Rahimi Department of Tissue Engineering and Applied Cell Sciences, School of Advanced Technologies in Medicine, Shahid Beheshti University of Medical Sciences, Tehran, Iran

Negar Ranjbaran Department of Genetics, Tehran Medical Sciences, Islamic Azad University, Tehran, Iran

Mostafa Rezaei-Tavirani Proteomics Research Center, Shahid Beheshti University of Medical Sciences, Tehran, Iran

Tannaz Safaralizadeh Cell Therapy and Regenerative Medicine Research Center, Endocrinology and Metabolism Molecular-Cellular Sciences Institute, Tehran University of Medical Sciences, Tehran, Iran

Masoumeh Sarvari Metabolomics and Genomics Research Center, Endocrinology and Metabolism Molecular-Cellular Sciences Institute, Tehran University of Medical Sciences, Tehran, Iran

Farshid Sefat Biomedical and Electrical Engineering Department, School of Engineering, University of Bradford, Bradford, UK

Interdisciplinary Research Centre in Polymer Science and Technology (IRC Polymer), University of Bradford, Bradford, UK

Motahareh Sheikh Hosseini Metabolomics and Genomics Research Center, Endocrinology and Metabolism Molecular-Cellular Sciences Institute, Tehran University of Medical Sciences, Tehran, Iran

Eda Sun Center for Stem Cell and Tissue Engineering Research & Practice, İstinye University, İstanbul, Turkey

Medical Faculty, Histology and Embryology Department, İstinye University, İstanbul, Turkey

Seyed Mohammad Tavangar Department of Pathology, Dr. Shariati Hospital, Tehran University of Medical Sciences, Tehran, Iran

Akram Tayanloo-Beik Cell Therapy and Regenerative Medicine Research Center, Endocrinology and Metabolism Molecular-Cellular Sciences Institute, Tehran University of Medical Sciences, Tehran, Iran

About the Editor

Babak Arjmand, M.D., Ph.D., was graduated from Iran University of Medical Sciences in medicine and received his Ph.D. in Applied Cell Science from Tehran University of Medical Sciences, Tehran, Iran.

He is currently Head and Director of Stem Cell and Regenerative Medicine Research Center, Endocrinology and Metabolism Cellular-Molecular Sciences Institute, Tehran University of Medical Sciences in Tehran, Iran.

His research activities concern the development of cell- and gene-based clinical products through the translational pathway from the basic to the clinic focusing on GLP, GMP, and GCP standards.

He has published more than 100 papers in journals and conferences proceedings.

He is the Member of different scientific committees and societies such as the following: Tissue Engineering and Regenerative Medicine International Society (TERMIS); Asia Pacific Association of Surgical Tissue Banks (APASTB); Iranian Tissue Engineering and Regenerative Medicine (ITERM); National Committee of Tissue, Cell, and Gene Therapy at Iran Food and Drug Administration (Iran-FDA); Iranian Council of Stem Cell Technologies; and National Working Group for Providing National Guideline on Stem Cell Therapy. He is also Editorial Board Member of several scientific journals in the field.

Trying to Reveal the Mysteries of Stem Cells Using "Omics" Strategies

Khadijeh Falahzadeh, Masumeh Jalalvand, Sepideh Alavi-Moghadam, Nikoo Bana, and Babak Negahdari

Abstract Owing to their self-renewal property and ability to differentiate toward various cell types of the human body, stem cells sound as an Utopian dream in regenerative medicine field. Thanks to remarkable achievements in novel technologies responsible for elucidation of stem cell biology, scientists try to revolutionize common concepts of conventional medications used for treatment of incurable diseases. Stem cells could ideally regenerate damaged tissues and eradicate the source of the problem and offer unique therapeutic strategies. Big data acquired by modern biological tools corresponding for different levels of governing stem cell fate help us more accurately identify fundamental principles involved in stem cell processes regulation. It has been revealed that key determinants of cellular behavior come from the five main omics studies encompassing genomics, epigenomics, transcriptomics, proteomics, and metabolomics. High-throughput analysis of multiple modalities of omics data contributes to precise understanding of sequential processes from DNA codes to metabolites which are the final products of decoding DNA information. Incorporation of these novel insights paves the way for the use of stem cells capacities in order to provide the most effective medications, drugs, as well as preventive health measure.

Genomics as the most mature omics bring genome-wide structural and functional data from the analytical evaluation of DNA. Epigenomics try to find the impact of the whole set of heritable chemical tags and modifications of chromatin

K. Falahzadeh
Department of Medical Biotechnology, School of Advanced Technologies in Medicine, Tehran University of Medical Sciences, Tehran, Iran

Metabolomics and Genomics Research Center, Endocrinology and Metabolism Molecular-Cellular Sciences Institute, Tehran University of Medical Sciences, Tehran, Iran

M. Jalalvand · B. Negahdari (✉)
Department of Medical Biotechnology, School of Advanced Technologies in Medicine, Tehran University of Medical Sciences, Tehran, Iran
e-mail: b-negahdari@sina.tums.ac.ir

S. Alavi-Moghadam · N. Bana
Cell Therapy and Regenerative Medicine Research Center, Endocrinology and Metabolism Molecular-Cellular Sciences Institute, Tehran University of Medical Sciences, Tehran, Iran

B. Arjmand (ed.), *Genomics, Proteomics, and Metabolomics*, Stem Cell Biology and Regenerative Medicine, https://doi.org/10.1007/978-3-030-27727-7_1

on cellular traits. Transcriptomics is the study of the complete set of RNA molecules within a cell at the given time and under specific conditions reflecting the level of RNA transcripts of the genome that are crucial for cell characteristics. The large-scale study of expressed proteins aims for the evaluation of the whole protein content of cells and finding their pivotal roles in the cell refer to proteomics. Metabolomics is defined as a comprehensive analytical study of small molecules considering as the final product of genome which represents the phenotype of each cell based on metabolite profile resulting from cellular metabolism.

Incorporation of multiple omics data leads to a deep understanding of underlying mechanisms regulating stem cell behavior and thus offers a full promise for introducing novel approaches for prevention, diagnosis, and treatment of numerous human diseases and specially the incurable one.

Keywords Stem cell · Genomics · Epigenomics · Transcriptomics · Proteomics · Metabolomics

Abbreviations

2D	Two-dimensional
2D-PAGE	Two-dimensional gel electrophoresis
3D	Three-dimensional
acetyl-CoA	Acetyl coenzyme A
A	Adenine
C	Cytosine
CAD complex	Carbamoyl-phosphate synthetase 2, aspartate transcarbamoylase, and dihydroorotase
CE	Capillary electrophoresis
CHD	Chromodomain helicase DNA-binding
ChIP	Chromatin immunoprecipitation
CID	Collision-induced dissociation
CRISPR/Cas9	Clustered regularly interspaced short palindromic repeats/clustered regularly interspaced short palindromic repeats associated protein 9
DBD	DNA-binding domain
DBS	Dried biofluid spot
ddATP	Dideoxyadenosine triphosphate
ddCTP	Dideoxycytidine triphosphate
ddGTP	Dideoxyguanosine triphosphate
ddTTP	Dideoxythymidine triphosphate
DNMT	DNA methyltransferase
DSB	Double stranded DNA break
ELISA	Enzyme-linked immunosorbent assay
ENCODE	Encyclopedia of DNA elements

ESC	Embryonic stem cell
ESI	Electrospray ionization
ETC	Electron transport chain
FADH2	Flavin adenine dinucleotide
FSC	Fetal stem cell
FT-IR	Fourier-transform infrared
FTMS	Fourier transform mass spectrometry
G	Guanine
GC	Gas chromatography
GIS	Gene identification signature
GLUT1	Glucose transporter type 1
GRN	Gene regulatory networks
GWAS	Genome wide association analysis
HAT	Histone acetyltransferase
HDAC	Histone deacetylase
HDM	Histone demethylase
HGP	Human genome project
HIF1α	Hypoxia-inducible factor-1α
HILIC	Hydrophilic interaction LC
HK	Hexokinase
HMDB	Human metabolome database
HMP	Human metabolome project
HMT	Histone methyltransferases
HPLC	High pressure liquid chromatography
HR-MAS	High-resolution magic-angle spinning
HSCs	Hematopoietic stem cells
IBD	Inflammatory bowel disease
ICM	Inner cell mass
IEC	Ion exchange chromatography
IMS	Imaging mass spectrometry
IMS	Ion mobility spectrometry
INO80	INOsitol requiring
iPSC	Induced pluripotent stem cells
ISWI	Imitation switch
K	Lysine
LC	Liquid chromatography
LDH	Lactate dehydrogenase
LDL	Low-density lipoprotein
LT-HSC	Long-term hematopoietic stem cell
m/z	Mass-to-charge ratio
MALDI	Matrix-assisted laser desorption/ionization
METLIN	METabolite LINk
miRNAs	MicroRNAs
MRI	Magnetic resonance imaging
mRNA	Messenger RNA

MRSI	Magnetic resonance spectroscopic imaging
MS	Mass spectrometry
MS/MS/MS2	Tandem mass spectrometry
MSC	Mesenchymal stem cell
NAD	Nicotinamide adenine dinucleotide
NADPH	Nicotinamide adenine dinucleotide phosphate
ncRNA	Non-coding RNA
NGS	Next generation sequencing
NIH	National institutes of health
NMR	Nuclear magnetic resonance
NP-LC	Normal-phase LC
NSC	Neural stem cell
OXPHOS	Oxidative phosphorylation
PAGE	Polyacrylamide gel
PAM	Protospacer adjacent motif
PC	Pyruvate carboxylase
PCR	Polymerase chain reaction
PDH	Pyruvate dehydrogenase
PDK	Pyruvate dehydrogenase kinase
PFK-1	Phosphofructokinase-1
pI	Isoelectric point
piRNA	Piwi-interacting RNA
PPP	Pentose phosphate pathway
PTM	Posttranslational modification
qPCR	Quantitative PCR
R	Arginine
RNA-Seq	RNA sequencing
ROS	Reactive oxygen species
RP-LC	Reversed-phase LC
rRNAs	Ribosomal RNAs
RT-PCR	Real time PCR
S	Serine
SDS-PAGE	Sodium dodecyl sulfate-polyacrylamide gel electrophoresis
SEC	Size exclusion chromatography
SMSC	Skeletal muscle stem cell
snoRNA	Small nucleolar RNA
SNP	Single nucleotide polymorphism
snRNA	Small nuclear RNA
SWI/SNF	Switching defective/sucrose nonfermenting
T	Thymine
TALEN	Transcription activator-like effector nucleases
TCA	Tricarboxylic acid
tRNA	Transfer RNA
UHPLC	Ultrahigh-pressure liquid chromatography
UV	Ultraviolet

WTSS	Whole transcriptome shotgun sequencing
Y	Tyrosine
ZFN	Zinc finger nucleases

1 Introduction

Stem cells with unique property to potentially generate almost all cell types of the body, as a powerful technology for understanding holistic molecular mechanisms governing cellular phenotype and behavior, shed light on stem cells clinical applications in regenerative medicine and developmental biology [1]. Regarding the recent decade achievements in stem cell field and post-genomic era it has been possible to propose new therapeutic strategies for a variety of incurable diseases [1, 2]. The integration of comprehensive and precise data acquired from multi-omics sciences encompassing genomics, transcriptomics, proteomics, and metabolomics, designated as systems biology, tries to elucidate all the underlying mechanisms involved in cell fate determination. These networks of cellular events in a molecular level with the aid of stem cell biology developments provide a versatile toolbox suitable for prevention, prognosis, diagnosis, and treatment of a wide range of incurable diseases. Moreover, multi-omics data analysis and molecular sequence annotations are conducive to clarify systems etiology of incidence of a variety of diseases [3, 4].

Integrating findings emerged from multi-omics investigations in stem cell biology area, as well as the most commonly used methods in each field will be discussed in detail in the subsequent sections of this chapter.

2 Genomics Basic Concepts in Stem Cell Area

Stem cells as undifferentiated cells are characterized by properties including self-renewal and ability to differentiate into different cell lineages [5]. They can be commonly split in accordance to their differentiation potency into four groups including totipotent, pluripotent, multipotent, and unipotent stem cells. Totipotent group can be differentiated into any cell types of the body including the embryonic and the extra-embryonic one [6, 7]. Other groups have less differentiation potency compared with totipotent type. Accordingly, pluripotent stem cells cannot produce extra-embryonic tissues and only can generate embryo-derived tissues [7, 8]. Multipotent stem cells are more committed and limited than totipotent and pluripotent stem cells, and they are capable to differentiate into defined lineages [9, 10]. The most committed group are unipotent stem cells that can exclusively differentiate into one particular cell type [11, 12]. Some examples of different types of stem cells include embryonic stem cells (ESCs), fetal stem cells (FSCs), induced pluripotent stem cells (iPSCs), and adult stem cells [12–14]. ESCs are obtained from the inner cell mass (ICM) of blastocysts. ESCs which display pluripotent potential are able to differentiate into three primary germ layers lineages including ectoderm, endoderm, and mesoderm [7, 15]. As the use of ESCs raised ethical problems, the introduction of alternative sources of stem cells including FSCs, adult stem cells, and iPSCs seems necessary. FSCs whose differentiation potential is lower than ESCs are multipotent cells that can be isolated from fetal tissues such as fetal blood, bone marrow, liver, and lung [16, 17].

Adult stem cells are multipotent stem cells and can be isolated from different adult tissues such as bone marrow, peripheral blood, cornea and retina of the eye, dental pulp of the tooth, liver, skin, gastrointestinal tract, and pancreas [18, 19]. Their self-renewal and differentiation capacity are more limited than PSCs and are considered as necessary factors required for tissue homeostasis. In addition, adult stem cells play an important role in repairing injured tissues due to their self-renewal characteristics [5].

Various types of stem cells with different origin have been found in adult tissues, providing valuable alternative sources for regenerative medicine [20]. Mesenchymal stem cells (MSCs) as multipotent stem cells can be originated from different adult and fetal tissues such as placenta, amniotic membrane and fluid, skin, adipose tissues, and peripheral blood. MSCs regarded as one of the most commonly used adult stem cells in cellular therapy [21, 22].

IPSCs are another group of stem cells that were created in the lab and show high similarity to ESCs, whereas they are differentiated adult cells which switch to ESC-like pluripotent stem cells by genetic reprogramming [20]. Although they don't have ethical issues like ESCs, more studies are needed to prove the safety of iPSCs for their clinical application in regenerative medicine area [23, 24]. Both types of pluripotent stem cells including ESCs and iPSCs are thought as pivotal tools to investigate tissue differentiation mechanisms, modeling of diseases, novel efficient drugs development, as well as tissue regeneration [25]. Recently the field of applied stem cells has been highly developed due to their potential to tissue regeneration. The study of stem cell biology at various levels of cellular fate regulation including genetic, epigenetic, transcriptomic, proteomic, and metabolomic is an important issue that among them genomics has been introduced as the most mature strategy [26].

Genomics as a well-known strategy is composed of different related disciplinarians such as genetics, molecular biology, biochemistry, statistics, and computational biology that focuses on the analysis of genome structure and functions. The structure of double helix DNA was introduced by Watson and Crick, then Sanger published the method to access DNA sequencing. Therefore, DNA was declared as the first sequenced molecule and the first organism that was sequenced was Haemophilus influenzae [27, 28].

In this regard, the human genome project (HGP) was an international project aimed for sequencing of the whole genome nucleotides and also characterizing genetic variations across the human genome [29–31]. At first, the project suggested in 1985 to identify the nucleotide sequences of the whole human genome. The HGP was launched in 1990 and supported by the National Institutes of Health (NIH) [31, 32]. The next international project after the HGP was the international HapMap project. HapMap tried to discover single nucleotide polymorphisms (SNPs) genetic variants [30, 33, 34]. The 1000 genomes project succeed to detect rare genetic variations among different populations using more advanced sequencing approach named next generation sequencing (NGS) and DNA arrays [30, 35].

Genome-wide association analysis (GWAS) as a successful procedure could classify many of the genetic variants associated with complex diseases among human populations. Studies on GWAS can help to figure out complicated phenotypes [36, 37]. NGS can be used for whole-genome sequencing and also exome sequencing [38–40].

In general, genomics can investigate genetic instructions in DNA, therefore merging stem cell biology and genomics can be considered as a novel approach to better understanding of stem cells behavior.

2.1 Genomics-Based Technologies

Despite the completion of the HGP, there are still many studies try to investigate and elucidate the functions of genes and the interactions between genes and their products. Since, changes in human genome can potentially lead to pathological conditions and threaten human health, the functional studies of biomolecules play a crucial role for finding the promising therapeutic strategies for a wide variety of diseases. There are different genetic methods that developed for studying genome, and each method has its own advantages and disadvantages [41]. Some of the frequently used techniques are briefly addressed here.

2.1.1 Gel Electrophoresis

This method is used to separate molecules based on their size and charge [42]. The process of DNA separation is conducted in an electric field. DNA with negative charge loaded into wells at cathode end of the gel with positive charge and via the force of electrical current moves through the gel in order to reach the anode end with positive charge. In this method DNA fragments with different sizes are separated in various gel concentrations with different pore sizes. There are two main types of gel used for DNA molecule separation including agarose gel and polyacrylamide gel (PAGE). Agarose gel is applied for pieces of DNA from 300 to 1000 base pairs while PAGE has smaller pores and is used for smaller DNA fragments with 10–500 base pairs in length [43].

2.1.2 Blotting

Blotting is a technique to transfer DNA, RNA, and protein molecules to a membrane sheet for determination of the presence or absence of a specific sequence among unknown fragments. Thus, there are several blot based on types of molecules [42]. Based on the molecule of interest which is aimed to be specifically detect, there are three types of blotting methods called Southern, Northern, and Western blotting for detection of DNA, RNA, and protein molecules, respectively. In Southern blotting method DNA fragments from an electrophoresis gel are transferred to a nylon or nitrocellulose membrane. Then, they are fixed and immobilized on the membrane. In immobilization process specific DNA fragments selectively hybridize to known labeled primers and visualize by autoradiography [44–46].

2.1.3 Polymerase Chain Reaction (PCR)

Polymerase chain reaction (PCR), also named conventional or end point PCR, is a chemical and enzymatic method that is generally used in molecular biology to amplify multiple copies of the given DNA segment. It can produce numerous copies of specific DNA from small amount of samples. The main reagents for PCR method are template DNA, two specific primers, nucleotides as building blocks for replication, and DNA polymerase as the key enzyme needed for replication process. There are different types of conventional PCR such as allele-specific PCR, asymmetric PCR, colony PCR, inverse PCR, multiplex PCR, and nested PCR. Each one become optimized which make them suitable for a particular purpose [47, 48].

PCR has a variety of applications in molecular biology and biotechnology branches such as gene therapy, genome manipulation, and biomedical areas. It has developed and provided fast and quantitative analyses of PCR by the advent of real time PCR (RT-PCR) or quantitative PCR (qPCR) [49]. QPCR is a more advanced type of PCR technology which can qualitatively examine genes expression level as well as defining the presence and amount of DNA product via comparing the proliferated DNA concentration with the housekeeping gene products, which supposed to show constant levels of the expression even in different cellular situations, and determine the quantity of target DNA [50].

2.1.4 DNA Microarray

The microarray is composed of a series of specific polymers, such as DNA, RNA, or cDNA, each one is attached to a specific probe on the surface of a chip [51]. Particularly, DNA microarray is used for detection of a large amount of target sequences at the same time. Oligonucleotides that attached as probes are labeled with fluorescent dyes and allow the formation of hydrogen bonds between target sequence and specific probes which are detected by emission of the fluorescent light. Tiling array is a subtype of microarray chips used for genome-wide functional analysis and detection of SNPs. Additionally, it is used for transcriptome mapping and protein-binding sites mapping of genome [42, 52, 53].

2.1.5 DNA Sequencing

Since the advent of DNA sequencing method in the 1970s a broad range of sequencing technologies were grown leading to improvement in speed and accuracy of sequencing methods. The Sanger method was the first-generation of DNA sequencing technique [54, 55]. The purpose of this technique is to determine nucleotide sequences in DNA fragments that is based on the replication of DNA in four separated reactions. Each reaction requires deoxynucleotides, adenine (A), thymine (T), cytosine (C), and guanine (G), as the building blocks for the newly synthesizing DNA strand, DNA polymerase as the key enzyme for polymerization of nucleotides, and labeled dideoxynucleotides including dideoxyadenosine triphosphate (ddATP), dideoxythymidine triphosphate (ddTTP), dideoxycytidine triphosphate (ddCTP), and dideoxyguanosine triphosphate (ddGTP) used for termination of replication process via incorporating to DNA strand which is being synthesized, and primers for the start point of polymerization. In each reaction, new nucleotides are added to new strand by DNA polymerase action that leads to DNA extension. If one of the dideoxynucleotides is added to the growing strand, the termination of replication is declared. Therefore, each reaction produces DNA fragments with different sizes that end with a specific nucleotide which includes each of ddNTPs. Ultimately, a large number of extended DNA fragments are produced in different lengths and then are denatured and separated based on their size by means of electrophoretic techniques. After that, DNA fragments are visualized via ultraviolet (UV) light or X-ray in order to read the sequences according to the termination sites. New advances in Sanger DNA sequencing technology led to automated and strategies using four ddNTPs, each labeled with a different color dye allows the high speed sequencing of DNA in one single tube [42, 56].

NGS as a high-throughput technique revolutionized the DNA sequencing technology. NGS by parallelizing the sequencing is able to create countless sequences. Interestingly, in comparison with traditional methods, NGS seems an inexpensive, more accurate, and faster method [57, 58].

2.1.6 Chromatin Immunoprecipitation (ChIP) Assay

In order to study the incorporation of a particular protein into specific DNA sequences, inside the native chromatin context of the cell, chromatin immunoprecipitation (ChIP) is used as a valuable method [59]. It can be applied to recognize various proteins correlated with a specific area of the genome, or multiple areas of the genome connected with a special protein [60]. The combination of ChIP with microarrays called ChIP-chip or with sequencing technology named ChIP-seq could be used to identify genome-wide discovery of the regains that are related to transcription factors and histone modifications [42, 61]. The high-throughput analysis of DNA-binding proteins and histone modifications contribute to understand the transcriptional control, epigenetic regulation, and genome decoding [62, 63].

2.2 Genome Editing for Stem Cell Engineering

Human pluripotent stem cells including ESCs and iPSCs sound promising strategies aimed for treatment of some chronic and incurable diseases and also powerful tools for disease modeling, drug discovery, and the study of stem cell biology [64, 65]. Genome editing is a newly developed technology in which genes of interest are specifically engineered and customized for either targeted therapy or genome manipulation that is achieved by the use of double stranded DNA break (DSB) via the action of endonucleases [66, 67].

Although gene therapy was previously considered as a strategy to add a new desired gene to human genome, it has been recently improved by development of new genome-editing technologies which mainly include zinc finger nuclease (ZFN), transcription activator-like effector nuclease (TALEN), and clustered regularly interspaced short palindromic repeats/clustered regularly interspaced short palindromic repeats associated protein 9 (CRISPR/Cas9). These methods as the most commonly used approaches in basic and clinical researches provide site specific, accurate, and efficient means for targeted addition, elimination, and alteration of given sequences at DNA level [68, 69].

In order to obtain more efficiency and accuracy in genome manipulation, ZFN, TALEN, and CRISPR/Cas9 technologies are noticed as powerful methods for human gene manipulation [70–72]. Since repair of DSB-induced gene is dependent on the endogenous DNA repair mechanisms, it can be generalized to any cell type or organism [68]. Hence, some platforms for inducing these site specific DSBs are briefly explained here.

2.2.1 Zinc Finger Nuclease (ZFN)

ZFNs are site-specific endonucleases generating by the fusion of a zinc finger DNA-binding domain (DBD) to a DNA-cleavage domain. Because zinc finger domains have been engineered to target specific DNA sequences, ZFNs are able to specifically edit the desired sequences of target sites. ZFNs are composed of several tandem zinc finger DBD coupled with the FokI endonuclease as their catalytic domain. The DBD itself contains three to six zinc fingers that detect the target sequence in order to break given nucleotides [67]. Although ZFNs were the first custom-engineered strategy used for human pluripotent stem cells genome editing, they are relatively complicated and challenging to be engineered which limit their wide applications. Regarding ZFNs design complexity, other alternative methods have been introduced for genome engineering [39, 67, 73].

2.2.2 Transcription Activator-like Effector Nuclease (TALEN)

TALENs have suggested another method for targeted genome editing arousing from the plant pathogen Xanthomonas [74]. The nuclease consists of an easily customized DBD fused with a nonspecific FokI nuclease domain. TALEN has been applied for producing gene reporters, knockout genes, and point mutations repair in human pluripotent stem cells [67, 75–77]. The appealing aspect of TALEN is the ease of manufacturing of novel engineered proteins and almost unlimited capacity for specific targeting of nearly any sequences [68, 78].

2.2.3 Clustered Regularly Interspaced Short Palindromic Repeats (CRISPR)

CRISPR as short palindromic repeated nucleotide sequences were originally discovered within the genome of bacteria and archaea which found to play a leading role in exogenous genetic elements removal and acting by the aide of some CRISPR-associated proteins or Cas proteins, mainly Cas9 protein [69, 79, 80].

Later on CRISPR/Cas emerged as the most frequently used genome editing strategy in various types of organisms. CRISPR/Cas9 system is composed of a type II CRISPR derived from bacterial immune system, RNA-guided DNA endonuclease, and Cas9 that act as an RNA-guided nuclease [68, 81]. The target site for CRISPR/Cas9 system incorporation on DNA is precisely selected by the sequence of guide RNA that is based on Watson–Crick base pairing [82, 83]. The guide RNA contains hybridizing part and the Cas9 interacting regions. These regions develop specific sites to hybridize Cas9 endonucleases [71, 82, 84]. The presence of a three-nucleotide protospacer adjacent motif (PAM) sequence composed of "NGG" at the downstream of the hybridization site is required for Cas9 association with PAM in the DNA target site [81, 82].

Among three types of CRISPR/Cas immune systems, type II CRISPR/Cas9 immune system sounds as the most powerful one for medical applications [85, 86]. A wide spectrum of RNAs could be designed and engineered intended for specific gene knockout, gene knock-in, and gene interference for targeted editing of given sequences of the genome [69, 78, 81].

Stem cells are an essential factor for tissue repair and regeneration of lost tissues. The CRIPSR/Cas9 has some advantages over the ZFN, TALEN in human pluripotent stem cells and somatic stem cells research. It has low cost, and in comparison ZNF and TALEN, CRIPSR/Cas9 is the fastest method [69].

The most prevalent application of mentioned genome editing approaches in stem cell area is the production of human disease models. With respect to fast growing iPSCs technology, it is feasible to manufacture patient-specific disease models in order to investigate various types of medication with optimal efficacy [87]. The sample isolation for patient-specific iPSCs has become available for being extracted from more easily achievable tissues like skin and blood [88, 89]. It is also possible to provide the normal and pathogenic or mutant types of cell lines for comparing different phenotypic traits between these two cell lines leading to the accurate investigation of mechanisms involved in manifestations of diseases [87].

It became possible to insert a desired mutation associated with a specific disease to generate the iPSC line of that disease for evaluating the effects of different treatments as well as the manipulation of gene sequence for the correction of mutant site [90, 91]. In this respect, genome editing strategies have been applied for healing some monogenic diseases by the use of patient specific iPSC lines [87].

Therefore, it has been proved that genome editing technics are powerful tools for ex vivo studies aimed for the genome manipulation. Gene therapy via patient-specific somatic cell and iPSC lines are safe and ready to be applied for therapeutic applications [92, 93].

Among ZFNs, TALENs, and CRISPR/Cas9 methods, CRISPR/Cas9 has been introduced as the most efficient and easily customizable method for generation of human site-specific gene modifications, iPSCs disease models, and animal disease models [39, 75, 92]. It is thought that CRISPR/Cas9 is potentially a standardized approach used for targeted and site-specific gene knockout, knock-in, and gene correction useful for diseases modeling via stem cell related approaches [94, 95].

The in vitro generation of organoids mimicking naïve organs like liver, pancreas, and stomach via adult stem cells has been reported [96–98]. These organoids supply a three-dimensional (3D) environment for modeling of human diseases and cancer therapy [99]. It appears that CRISPR/Cas9 could act as a powerful tool for genome engineering and editing of human genome in adult stem cells originated organoids [100]. Diseases modeling allow the in-depth investigations of pathogenesis of diseases for introducing novel therapeutic approaches [101].

2.3 Epigenomic Aspects of Cell Fate Governing

Epigenetic is described as any changes leading to inheritable alteration in genome function, without any effect on DNA sequences [102, 103]. Although the protein expression profiles are crucial factors to determine the characteristics of the cell structure and functions according to genomic sequence, epigenome and its regulatory mechanisms with leading role in cell fate governing seem an appealing field to disclose gene expression regulation determinants [102, 104]. For instance, the comparison between gene expression profiles of differentiated cells and ESCs shows that ESCs genome is transcriptionally more active than somatic cells [105, 106].

The development of stem cells like any other cells is associated with changes in cellular transcription state that lead to activation or silencing of specific genes. After differentiation induction of pluripotent stem cells, some multipotent progenitor genes have to be activated while stemness-related genes have to be inactivated which are mainly controlled via cytokines, growth factors, morphogens, and cofactors [106–108].

Therefore, during transition a stem cell to a committed cell, progressive gene silencing reduces the multipotency of stem cells [109]. Stem cells react to external signals, therefore their intrinsic state can be altered and these changes could impact their developmental potential and lead to more limited potency than their previous statues [109–111].

Some common covalent modifications of chromatin include DNA modifications like DNA methylation and histone modifications such as methylation, acetylation, and phosphorylation. These reversible modifications are essential for controlling the cellular response to internal and external stimuli in different condition which govern the cell behavior [26, 112].

Chromatin modifications chemical alterations affect chromatin states and regulate the balance between euchromatin and heterochromatin that in turn determine the transcriptionally active or silent regions of genome by specific enzymes, ATP-dependent chromatin remodelers, and incorporation of histone variants [113–115]. Main cellular mechanisms regulating cell fate determination such as DNA replication, transcription, and also DNA repair can be affected by epigenetic alterations [116].

Hence, given the importance of the epigenetic alterations in determining the fate of the cell and developmental processes, some of the notable modifications including DNA and histone methylation, histone acetylation and deacetylation, and chromatin remodeling factors are described below.

2.3.1 DNA Methylation

DNA methylation was identified by chromatography-based techniques about 70 years ago. It occurs in the context of CpG dinucleotide at the fifth position of the cytosine carbon which is followed by guanine nucleotide catalyzed by a family of DNA methyltransferases (DNMTs) [116, 117]. The addition of methyl group on CpG in the promoter regions generally results in the transcription repression of given genes and in mammalian is commonly caused by the action of DNMT1, DNMT3a, and DNMT3b [104].

2.3.2 Histone Modifications

Histone chemical epigenetic marks are mainly caused by enzymes including histone acetyltransferases (HATs), histone deacetylases (HDACs), histone methyltransferases (HMTs), and histone demethylases (HDMs) [116, 118]. These posttranslational modifications (PTMs) can be added to several different types of amino acid residues on histone proteins including lysine (K), serine (S), arginine (R), and tyrosine (Y), although K is the most predominant one [119].

Histone methylation is occurred by the action of HMTs and mostly caused heterochromatin sites commonly at centromeres and transposons [106, 120]. Their modification is more complicated than DNA methylation because of the recruitment of different modifier enzymes [113].

By adding the acetyl group HAT can change the charge of the amino group in lysine residue at histone tails which mainly result in activation and upregulation of given genes, while HDACs can remove acetyl group at histone tails and inhibit gene transcription [113, 121].

2.3.3 Chromatin Remodeling Factors

Chromatin remodeling factors allow the removal or addition of nucleosomes in specific regions of the genome and cause transcriptional activation or suppression, respectively [122]. These structural changes are done by recruitment of ATP-dependent

Table 1 The list of some common epigenetic marks on histones and their impact on transcription. The most important amino acid on histone tail is K that undergo different covalent modifications resulted in transcriptional repression or activation

Histone modifications	Example	Common effect
Methylation	H3K27me3	Transcriptional repression
	H3K9me3	Transcriptional repression
	R8me	Transcriptional repression
	H3K4me3	Transcriptional activation
	R43me	Transcriptional activation
	H4K3me3	Transcriptional activation
Acetylation	H3K14ac	Transcriptional activation
	H3K4ac	Transcriptional activation
	H3K9ac	Transcriptional activation
	K18ac	Transcriptional activation
	K23ac	Transcriptional activation

chromatin remodeling complexes such as switching defective/sucrose nonfermenting (SWI/SNF), imitation switch (ISWI), chromodomain helicase DNA-binding (CHD), and inositol requiring (INO80) in mammalian [123–125]. Some well-studied examples of human epigenetic marks relevant to transcriptional repression or activation, discovered by ChIP-based methods, are listed in Table 1 [119].

Histone modifications or histone codes play pivotal roles in stem cells self-renewal maintenance or trigger for differentiation via downregulation or upregulation of specific genes. Based on reported histone codes, stem cells are supposed to show transcriptional activation marks at promoter regions of genes which are responsible for stemness statues and transcriptional repression of genes associated with differentiation process [109]. Although this is a generally approved rule, some bivalent histone modifications changed this straight forward concept. H3K4me3 and H3K27me3, for instance, are the most prominent bivalent marks in ESCs. Bivalent modifications are abundant in ESCs promoter regions named "poised promoters" and make them ready for immediate response to stimulation or differentiation induction signals in spite of being functionally inactive [126–128].

Other bivalent domains in ESCs are H4K3me3 and H3K27me3 which are proposed as transcriptionally active and silent epigenetic marks, respectively. The simultaneous detection of these modifications at promoter regions of transcription factors that are components of regulatory complexes involved in differentiation and lineage specifications provides other evidence for poised promoters function [127].

The association between epigenetics and stem cells came from "epigenetic landscape" suggested by Waddington in 1900s. He claimed that increased level of methylation parallel with decreased level of acetylation lead to specialization from pluripotent stem cells to different cell lineages [129]. Due to influential role of epigenetic marks on gene expression profiles of stem cells, it is not surprising that epigenetics seems as an effective strategy for stem cell manipulation in order to produce specific cell types, especially in case of cell reprogramming and transdifferentiation, for regenerative medicine goals [129, 130].

Epigenetic modifications have been found as an important factor for iPSCs generation and therefore reprograming of differentiated cells. It has been demonstrated that through somatic cell reprograming with different origins the number and sites of methylation marks specifically differ from the ancestral somatic cells [131]. Moreover, a comparative study attended for detection of DNA methylation status in different ESCs, iPSCs, as well as somatic cell lines revealed that substantial differences between these cells are attributed to only 10% difference in DNA methylation site which highlight the prominent role of epigenetics in stem cells fate control [132].

2.4 Stem Cells Transcriptome

DNA is the source of genetic information in the cell which is decoded via genes transcription process. RNA molecules with structural and regulatory functions are mediators between DNA and proteins. The whole set of RNAs in a biological unit like a cell is named transcriptome and its study is called transcriptomics [133]. The first stage of gene expression regulation is transcription in which DNA is transcribed into different types of RNAs [134].

Transcriptomics approaches try to qualitatively measure the level of RNA expression in the given genome. In the past decade, transcriptomics has revealed that despite the fact that approximately up to 80% of the human genome is transcribed, only about 3% of those transcribed RNAs undergo translation process and encode proteins [26, 135]. In general, transcriptomics measures the level of RNA from aspects qualitative such as recognizing new splice and RNA editing sites, and quantitative such as measuring the level of expression of the transcript in the genome [26].

Based on protein coding capacity of RNA molecules they are classified into two major groups as follow coding and non-coding RNAs. The messenger RNA (mRNA) molecules as the coding RNAs are translated into proteins while non-coding RNA (ncRNA) molecules could not produce protein or are poorly translated. A large number of ncRNAs including ribosomal RNAs (rRNAs) and transfer RNAs (tRNAs) are associated with mRNA translation process, small nuclear RNA (snRNAs) are involved in splicing of introns from primary transcripts, and small nucleolar RNAs (snoRNAs) contributed to the modification of other RNA types, predominately. NcRNAs show various functions. A large number of ncRNAs including rRNAs and tRNAs are associated with mRNA translation process, snRNAs are involved in splicing of introns from primary transcripts, and snoRNAs predominately contribute to the modification of other RNA types [136, 137].

MicroRNAs (miRNAs) are an important class of ncRNAs with evolutionary conserved sequence that play a key role in regulation of gene expression via specific binding to target mRNAs and consequence suppression of protein production. Piwi-interacting RNAs (piRNAs) that form complexes with piwi proteins are involved in posttranscriptional regulator and seem as epigenetic regulators of gene expression [136, 138].

The first study in transcriptomics area was launched in the early 1990s. Investigation of gene expression in different status and time point enables better understanding of detailed gene expression process and relevant regulation mechanisms. In order to study changes during gene expression in various organisms, transcriptome analysis is employed that highlight its important role for elucidation of molecular pathways involved in human diseases [133].

Beside these fundamental roles of transciptomics, it is applied for uncovering the stem cell regulatory mechanisms which finally control the functional and structural characteristics of stem cells. Transcriptomics enable us to distinguish cell states in cell populations, and to investigate the heterogeneity at high resolution. Therefore, transcriptomics is regarded as a powerful tool to reveal gene regulatory networks during differentiation and stemness maintenance, and molecular mechanisms involved in the processes followed by induction of somatic cells to iPSCs [139, 140]. Gene regulatory networks (GRNs) are fundamental regulators for pluripotency maintenance by the aid of a variety of transcription factors that paved the way for describing regulatory mechanisms involved in stem cell behavior etiology [140].

Hence, transcriptomics could provide a new vision of pluripotent stem cells to understand gene regulation and regulatory networks during differentiation, and molecular mechanisms during turning somatic cells into the induced pluripotent stage [139].

During differentiation of pluripotent stem cells, the expression of some transcription factors related to pluripotency is reduced [139]. Therefore, studying these mechanisms, that are important to maintain self-renewal and differentiation potency, are required to distinguish the exact features of pluripotent stem cells transcriptome [141].

Microarrays are composed of short labeled oligonucleotides or probes that represent another valuable method used in transcriptomics. Microarrays enable the measurement of the frequency of specific transcripts which are hybridized to an array of given probes. The intensity of fluorescent emission of labeled probes after hybridization is representative of the frequency of a large amount of transcripts simultaneously. This technic at the same time can assay a lot of transcripts at low cost [133, 142, 143].

In order to interpret the stem cells transcriptome, new achievements in traditional methods such as cDNA sequencing and gene identification signature (GIS) have been applied. These techniques suffer from high cost and being low-throughput. By the advent of RNA sequencing (RNA-Seq) technology in 2006 as a fast and inexpensive approach, accurate transcriptome roles clarification for stem cells and any other cell types have been facilitated [133, 144, 145].

RNA-Seq could be used as a whole transcriptome shotgun sequencing (WTSS) technic intended for RNA sequencing that can detect the abundance of RNAs through NGS [146]. RNA-Seq has advantages over other techniques since it gives more information about the expression profile of given genes, allele-specific genes, and also alternative splicing that are essential factors needed for gene expression regulation [136].

Transcriptomics is thought as a reliable strategy to compare differentiation potential of various types of stem cells with different origins. Richards et al., for instance, used proteome profile analysis of two human ESC lines and contrasted them with the murine ESCs. They introduce some markers of human ESC lines associated with their stemness phenotype [147]. The comprehensive proteome profiling of pluripotent stem cells is required to annotate marker genes functions that contribute to introduce novel genes and transcripts with significant impacts on pluripotent stem cells gene regulations. In this regard, the encyclopedia of DNA elements (ENCODE) project and also Au's study try to elucidate the features of human ESCs whole transcriptome profiling [141].

Another study for finding proteome differences between distinct stem cell types has compared the gene expression profile between human pluripotent stem cells-derived cardiomyocytes and the fetal one. They revealed that after isolation of these pluripotent stem cells-derived cardiomyocytes they resemble the first trimester fetal heart cells. Subsequent culture and differentiation induction process altered human pluripotent-derived cardiomyocytes proteome profile in a way that they exhibit properties like second trimester heart cells [148].

Finding the mechanisms involved in differentiation of different cell types from the same genome via gene expression regulation is a fundamental challenge in stem cell biology. Thus, transcriptome analysis is a valuable tool to assess gene expression changes and to explore stem cell fate regulatory factors [133].

3 Proteomics Insight into Cell Biology

Proteins as crucial bioactive elements can regulate signaling pathways via specific mechanisms to form molecular information of the cells. Additionally, the cellular biological functions can be affected by various types of proteins [149–151].

Overlay, the whole protein content that are expressed by cells in a certain time are called proteome. Study on proteome can show the percent of dynamic proteins in multicellular organisms and also their correlations with different diseases [152–155]. The large-scale study of proteome, which can identify and quantify the proteome and proteins localization, interactions, and posttranslational modifications, is called proteomics [156–158]. Regarding technical matter, based on the high dynamic range and limitations of low abundance proteins evaluation, the loss of effective purification, and identification methods, the study of proteome is really a challenging issue [159].

For the first time, "proteome" and "proteomics" terms were used by Wilkins in 1994 [152, 160]. Proteome data included general cellular programs, cell cycle, and energy metabolism reflect the cell identification. In recent years, advancements in proteomics technologies widen the applications of proteomics for understanding molecular interactions and collaborations as the root of cellular structure and functions, remarkably in stem cell biology [161, 162].

Cell proteins dynamically change in response to environmental stimulus. Therefore, recognizing the dynamic interactions between proteins and other biomolecules helps to recognize the biological mechanisms and biochemical reactions take part in cellular processes [163]. Accordingly, proteomics strategies can provide a precious point of view to the molecular events participate in self-renewal, proliferation, and differentiation procedures of stem cells [156, 158]. On the other hand, scientists can characterize a subset of differentiation-specific proteins which can be used as hallmarks for the revealing stem cell behavior via proteomics-based tools [164, 165].

The ability to quantify proteins in the various cell populations can lead to proteomics progression into the novel pathways governing cell fate [166]. Furthermore, proteomics can be called "post-genomic" science, because it plays a significant role in passing genomics to clinical applications, especially in the field of diseases prevention and diagnosis. Proteomics is much more complex than genomics, mainly because the genome is constant, but proteome differs from cell to cell due to its dynamic feature. Proteomics seems as a rapidly advancing field of science in post-genomic area, which is significantly important to uncover cellular functions. Moreover, proteomics analysis approaches can employ multiple techniques to identify the expression, validation, and quantification of specific proteins [167–169]. Of course, as a relatively newly developed branch of science, some proteomic techniques need further optimizations in terms of materials, tool design, sensitivity, and efficacy [170].

3.1 An Overview to Proteomics Approaches

Proteomics approaches contain a wide range of techniques from conventional to advanced one. Some commonly used proteomics methods include chromatography-based technologies, enzyme-linked immunosorbent assay (ELISA), Western blotting, protein sequencing, protein microarray, gel-based strategies, X-ray crystallography, mass spectrometry (MS), and nuclear magnetic resonance (NMR) spectroscopy techniques [171–174].

3.1.1 Chromatography-Based Methods

Chromatography-based techniques refer to a series of separation methods in which molecules are separated in two phases called stationary and mobile phases. The stationary phase of the chromatography is a porous substrate which placed in the column and the mobile phase is a washing solution that is pumped in above the column. As a general rule, molecules which tend to stationary phase are moving slower than those tend to the mobile phase [175–177].

According to these phases, two major classes of chromatographic methods are introduced as follow: liquid chromatography (LC) and gas chromatography (GC) are common types of chromatography-based methods in which the mobile phase is liquid and gas, respectively. In GC, isolation and investigation of the volatile matters are done without their decomposing with generally higher resolution than LC. Among various LCs the most efficient types are called high-performance liquid chromatography or high pressure liquid chromatography (HPLC) and ultrahigh-pressure liquid chromatography (UHPLC), which are extremely high-speed and high-resolution methods that can apply high force to press sample within the column [178–180].

Ion exchange chromatography (IEC), size exclusion chromatography (SEC), and affinity chromatography are some prevalent chromatographic techniques. IEC can purify proteins based on their surface charged groups. Indeed, proteins have different amino acids sequences. Some amino acids are cationic and some others are anionic. In physiological pH, the net charge of the protein is assessed by the balance between these charges. Polymer resin with covalently affiliated operative ionic groups is used as the IEC stationary phase. These stationary phases are arranged in various exchange groups due to their acid/base features. Base exchange includes positive charge groups that are called anion exchangers. Cationic exchange carries negative charge groups that absorb positive charge groups. Generally, IEC purification is based on a reversible interaction between a protein and a charged substrate or ligand with opposite charge [152, 181–183].

In SEC or molecular sieve chromatography, proteins are purified according to their molecular size [184, 185]. The interaction between particular ligand coupled with a chromatographic matrix is the basic principle of the affinity chromatography. Investigators can evaluate protein degradation, post-translational modifications, and protein–protein interactions via affinity chromatography [149, 186].

3.1.2 Enzyme-Linked Immunosorbent Assay (ELISA)

The basis of the ELISA test as a highly sensitive immunoassay is on the response of antibody to the antigen. In this test, a specific marker conjugated with antibody reacts with a specific antigen and finally, the concentration of antibody and antigen can be assessed by a spectrophotometer. In this assay the intensity of colors is representative of the concentration of antigen in the sample [149, 187, 188].

3.1.3 Western Blotting

Western blotting which also known as immuno-blotting or protein-blotting is one of the key techniques in cellular and molecular biology that is commonly used to detect the presence of a desired protein among a series of cellular proteins. This method is more specific than ELISA test but is less sensitive. Because of its high cost, Western blotting is not the first choice for detection of proteins in a mixture of different biomolecules and is more likely to be used to confirm the positive results of the ELISA test [45, 189].

3.1.4 Sequencing Methods

Determining process of the whole or some part of amino acid sequences in a given protein is called protein sequencing which can characterize both amino acid sequence and post-translational modifications of the protein. The two main systems for protein sequencing are MS and Edman sequencing method. Edman sequencing can ascertain the amino acid sequence in peptides or proteins and includes chemical reactions which can sequence the N-terminus amino acid of the protein and then allow the identification of protein through searching this sequence in the protein databases [190, 191].

3.1.5 Protein Microarrays

Protein microarrays or protein chips are a group of proteomics techniques that contain analytical protein, functional protein, and reverse-phase protein microarrays. Antibody microarray as the most typical class of analytical protein microarray can detect proteins by means of direct protein labeling and measure the binding affinities and expression levels of proteins. Functional protein microarray provides the condition for the study of multiple protein interactions such as protein–protein, protein–DNA, protein–RNA, protein–lipid, protein–drug, and etc. Reverse-phase protein microarray is applied for detecting the modified and dysfunctional protein biomarkers of a particular disease. Commonly, protein chips as a high-throughput advanced technique suggest a solution for analyzing low-abundance proteins in the cell [171, 192, 193].

3.1.6 Gel-Based Methods

Sodium dodecyl sulfate-polyacrylamide gel electrophoresis (SDS-PAGE) and two-dimensional gel electrophoresis (2D-PAGE) are the main widely used gel-based proteomics approaches. In SDS-PAGE method that is used for protein separation, after adding SDS that is an anionic detergent the protein is linearized with equal charge. Then, proteins are separated in a medium based on their molecular weight by moving through an electric field in a way that small proteins move faster than the large one. Protein standards with defined sizes are used to recognize the unknown protein in a sample. The principle of protein separation in 2D-PAGE is based on two characteristics of protein including mass and charge. In the first dimension proteins are separated according to their isoelectric point (pI), the pH in which biomolecules net electrical charge is zero. In pI value, proteins are no longer move through the gel and are ready to be separated in the second dimension based on their molecular weight. Separated proteins are stained and visualized as distinct spots that can discriminate proteins with similar molecular weights. 2D-PAGE considered as a useful method for post-translational modifications, mutant protein characterization, and protein identification in complex protein mixtures [152, 194–197].

3.1.7 X-Ray Crystallography

3D structure identification of proteins can be provided by X-ray crystallography. In this method, crystallized samples are exposed to X-ray and the following diffraction patterns are analyzed in order to interpretation of the data about the size of the repeating part which forms the crystal and crystal packing arrangement. Some of the X-ray crystallography applications are: the study of the protein-nucleic acid complexes, virus systems, site-directed mutagenesis, illumination of enzyme mechanisms, drug designing, and protein–ligand interactions [198, 199].

3.1.8 Spectroscopic Methods

MS and NMR spectroscopy suggested as two high-throughput proteomics methods which are reviewed in the Sect. 3.2 [200].

3.1.9 Bioinformatics

Bioinformatics is required for proteomics analyzing in order to manage enormous and heterogeneous proteomics data. It is advanced parallel with the development of high-throughput technologies that need powerful data analysis that there are limitations around the management of a huge amount of data. Additionally, finding the association between proteomic data and the other omics technologies can be facilitated by the application of relevant databases which finally allow accuracy determination of networks of data come from different methodologies [201, 202].

3.2 Proteomics Applications in Stem Cell Area

Controlling the fate of stem cells is completely depends on the interactions between stem cells and their niche including the signals from adjacent cells, extracellular matrix, and pathogens. Therefore, understanding of the protein-protein interactions mechanisms in a cell and also uncovering the mechanisms responsible for cell replying to intrinsic and environmental signals are influential factors to our perception of proteomics. In this regard, applying proteomic approaches for the study of several cell populations, investigating the cellular proteins incorporations, and their functional mechanisms can be efficient. The high-throughput proteomics technology allows uncovering signaling pathways, protein ontology, protein–protein interactions, and protein–DNA interactions [20, 203, 204].

Analysis of protein localization, important for clarifying protein functions, is a significant regulatory point for stem cells fate determination that becomes feasible via proteomics. It has been proved that the mislocalization of proteins in a cell can cause various human disorders. For instance, alternation in protein localization of

ESCs can turn their self-renewal maintenance function into differentiation [156, 205, 206]. Hence, analyzing molecular mechanisms and signaling pathways related to the maintenance of ESCs in the undifferentiated state, regulation of their differentiation, and pluripotency will be required to know how ESCs can be applicable in cell therapy and regenerative medicine. Consequently, proteomics approaches may play a significant role in this area [207].

Pluripotency of iPSCs has confirmed by some tests containing teratoma formation, in vitro differentiation, tetraploid complementation, and chimera formation. Despite that, some notable variations between iPSCs and natural ESCs may be observed. Therefore, molecular characterization and proteomic evaluation of stem cells are required to examine the diversity and correlations between these two cells before translating to clinical setting [65, 208, 209].

The therapeutic efficacy of stem cell-based regenerative medicine treatments is attributed to their differentiation capacity and engraftment ability. Hence, it is a matter of great importance to reveal the biological mechanisms related to events defining stem cell properties. Furthermore, recent proteomic analysis of cell secretome and physiological conditions can be provided via the advanced analytical proteomic techniques [210, 211]. As a result, proteomics technology can offer opportunities for the development of novel and effective treatments for incurable diseases through the evaluation of stem cell proteome. In other word, proteomics incorporates the sum of high-throughput techniques to identify proteins which play the key roles in the cell physiological processes. Hence, this technology is an applicable tool to uncover what is crucial for promotion of cell therapy and regenerative medicine [165, 212].

Despite the technical advancement of stem cells transplantation, there are still some barriers for stem cell therapy like approving the safety and quality of stem cells for treatment that hinder the fast development of stem cells therapies. In this regard, proteomic evaluation of given cells and tissue biomarkers after and before cell therapy could be an effective solvent. In fact, it is possible to compare the proteins present in healthy people with patients who undergoing stem cells therapy for evaluation of the efficacy of the treatment. Therein, pharmacoproteomics is a high-throughput proteomics platform which is worthy in therapeutic biomarker identification and plays a chief role in personalizing medicine [156, 212].

It could be suggested that proteomics progression has a great promise to the evolution of personalized medicine through promoting the proteomics-based molecular diagnostics, and detection of biomarkers. To this end, personalized medicine is a novel field of medicine which tries to offer a customized healing procedure for each individual based on their proteome and metabolome characteristics and according to the conditions of their cellular physiology, environment, and medical history [154, 213, 214].

Generally, proteome analysis in regenerative medicine and personalized medicine area gives the perfect view of cellular information using different proteomics and metabolomics techniques with the aide of bioinformatics databases for interpretation of acquired data. These massive data provide promising strategies for treatment of incurable diseases (Fig. 1).

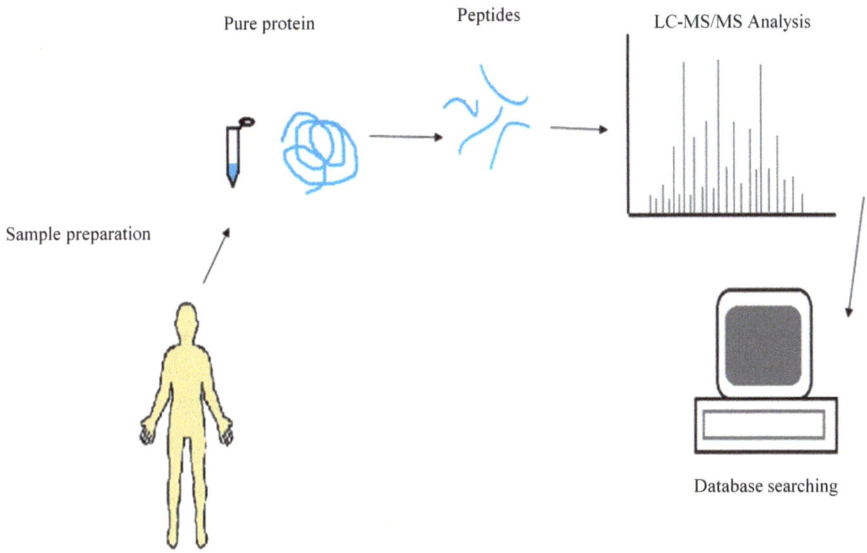

Fig. 1 Overview of proteomics analysis steps. After protein sample fractionation, cell proteins can be digested to form peptides in order to be analyzed by computational proteomics approaches like LC-MS. Following that, acquired data are compared with databases to be interpreted

4 Metabolomics and Fundamental Biological Processes

Metabolomics refer to the systemic and comprehensive identification and quantification of small molecules, metabolites, and their modifications in a complex mixture [215, 216]. These low molecular weight metabolites include carbohydrates, amino acids, fatty acids, nucleotides, organic acids, ions, etc. are formed by anabolic and catabolic networks of the cell [217]. Due to the fact that metabolites as the final products of genome are accurately assessed via metabolomics approaches, metabolomics is supposed to be the last part of omics chain indicating the phenotype of the organism [218].

Metabolomics could be considered as a multidisciplinary technology that introduces versatile approaches to measure metabolic alterations after induction of different agents like environmental and pathological factors. These metabolic changes caused by cell response to various stimulations raised from alteration in metabolic pathways and biochemical interactions governing cellular process in different situations [217]. So far, metabolomics is thought to be the most powerful set of techniques intended for metabolites evaluation in a given sample which could be biofluids or tissues at a certain point in time [219].

The human metabolome project (HMP) was launched in 2005, aimed for assisting scientists to reveal the metabolites involved in cellular process and uncovering the association between metabolites and metabolic disorders. Metabolomics outstanding achievements depend on progression of new tools and approaches which allow the high-throughput assessment of metabolites participate in cellular process [217].

The most widely analytical methods used in metabolomics studies are NMR, MS, and Fourier-transform infrared (FT-IR) leading to accurate and sensitive of downstream gene products measurement, metabolites identification and qualification, biomarker discovery, efficacy evaluation of therapeutic methods, drug metabolism exploring, and toxicity assessment [217, 220].

Systems biology via using computational models integrates the metabolites data through different scales such as cells, tissues, or a whole organism with physiological functions [221]. It contributes to metabolomics studies by discovering biochemical reactions in complex organisms which holistically provide a top-down integrated view of biological systems [222]. With the aid of advanced software and analytical approaches systems biology plays a fundamental role in metabolomics area by introducing novel systems-level insights into explaining interactions and dynamics of complex organism at different scales [221, 223].

To elucidate the mystery of biological processes and to provide a detailed analysis of interaction networks in a living organism, it is highly recommended to study their structural and functional properties in a real situation with a variety of interactions between different parts of a cell or organism instead of their study as separated parts of organism, as if possible [221, 223].

The advancements of chromatographic-based technologies led to the emergence of metabolic profile term by Horning in 1971 [224, 225]. Robinson and Pauling are thought as pioneers of metabolomics studies in 1971, although they did not clearly name their study as metabolomics one [226].

The term metabolome was first used by Oliver in literature in 1998 [227]. The metabolomics term was first coined by Nicholson et al. in 1999 in literature [228]. Following rapid instrumental developments in spectroscopic approaches, Nicholson was the first one who introduced NMR spectroscopy as a valuable tool for pattern recognition of analytes in biological samples [229, 230].

Five years later, the first metabolomics society was established in order to enhance our knowledge of metabolomics in the life sciences in 2004 [218]. Increased attention to metabolomics led to development of web databases including METabolite LINk (METLIN) and the human metabolome database (HMDB). METLIN was released in 2005 as an open access database which contains the structural information of metabolites based on Fourier transform mass spectrometry (FTMS) spectra, tandem mass spectrometry (MS/MS) spectra, and LC/MS analysis [231]. HMDB is the most comprehensive database in metabolomics area which includes the largest databank providing the concentrations of organism-specific metabolites, small molecules related to biochemical pathways, and cancer biomarkers. The quantified information in HMDB comes from MS and NMR experiments via analyzing a variety of samples such as urine and blood [232, 233].

Since then, metabolome-wide association studies have attracted a lot of attention aimed for uncovering metabolic pathways and biochemical reactions governing structural and functional features of a cell, tissue, or organism. Two main categories of metabolomics are targeted and untargeted metabolomics. Targeted metabolomics is a powerful strategy for measurement and quantification of defined groups of metabolites in a sample with more sensitivity and specificity than untargeted method. Owing to the fact that in targeted technology limited numbers of metabolites are under investigation, the final peaks of metabolites are less complex than untargeted one leading to more simple data interpretation. Untargeted studies refer to comprehensive analysis for identification and quantification of unknown analytes or their unexpected changes in a complex mixture thereby spectra derived from MS are compared with databases in order to discover the identity of given metabolites. Untargeted metabolomics is thought as a valuable tool for biomarker discovery in biological samples [234, 235].

4.1 Metabolomics: An Emerging Technology for Basic and Medical Sciences

Metabolomics as an emerging field of science is supposed to revolutionize the basic, experimental, and medical biosciences by identification and quantification of unknown or given metabolites which clearly interpret the phenotypic traits and the cellular biochemical pathways. Impressive achievements in spectroscopic techniques and NMR spectroscopy allow the wide range of metabolomics-based applications such as biomarker and drug discovery specially for targeted therapy of incurable diseases, drug resistance mechanisms, developmental process investigation, assessment of differentiation and self-renewal status of stem cells, evaluation of the organism or cell response to therapies, pathological agents or stimulatory factors, decoding variations of genome and proteome products for each individual, elucidation of metabolic changes and biochemical signature of different diseases like cancers [236–239].

High-throughput platforms obtained from metabolomics strategies are suggested as a way to validate the direct genomics, transcriptomics, and proteomics evidence that could potentially explain various aspects of biomolecules functions (Fig. 2) [217, 237].

Some well-known critical metabolites discovered as biomarkers encompass acylcarnitines for fatty acid oxidation disorders, creatinine for renal function, low-density lipoprotein (LDL) and cholesterol for cardiovascular diseases, hippuric acid for liver diseases, histamine for allergies and mast cell disorders, and citric acid for metabolic disorders. These metabolites are specific indicators that allow the prediction of relevant diseases. Additionally, glutaric acid, hydroxyglutaric acid, methylsuccinic acid, ethylmalonic acid, and alloisoleucine are suggested as foremost factors to newborn screening which hold great promise for novel applications of targeted metabolomics [216].

The biological specimen or primary sources for analytical approaches in metabolomics are mainly cells, tissues, and biofluids such as urine, blood, serum, and plasma. The mentioned biofluids are considered as the most frequently used samples due to the fact that their isolation and preparation are relatively easy [240].

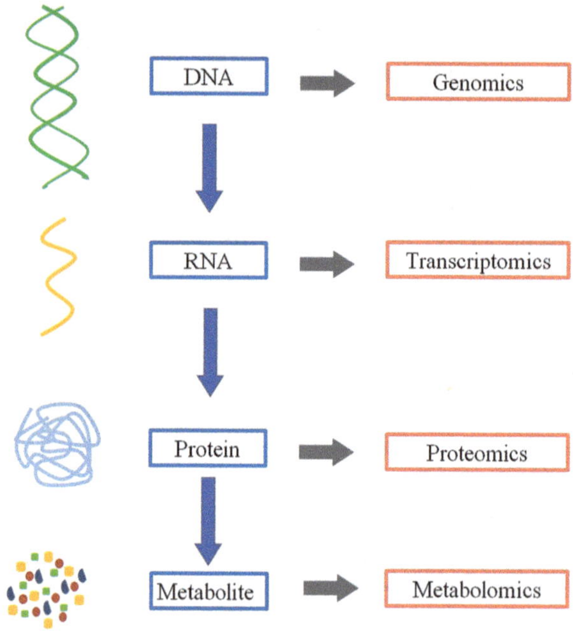

Fig. 2 Four main levels of biological investigation including genomics, transcriptomics, proteomics, and metabolomics. Integrated studies using high-throughput multi-omics approaches are thought as the most accurate tool for discovering molecular events intended for understanding mechanisms involved in cell structure and functions governing

Some other sources for metabolites analysis could be saliva, cerebrospinal or amniotic fluid, dried biofluid spot (DBS), semen, bronchoalveolar lavage, sweat, and stool. Interestingly, intact tissue could be used as the primary sources for metabolomics studies enabling scientists to detect and quantify metabolites involved in a specific disorder. This technology provides more reliable and accurate information than isolated samples [216].

Several sample types utilized for identification of thousands of metabolites in the body leading to phenotypic annotations. By discovering biomarkers associated with specific diseases metabolomics could be introduced as a multi-purpose tool for prognosis, diagnosis, and treatment of a wide range of diseases [241–243].

The exogenous and endogenous factors including age, gender, eating habits, ethnic background, drug metabolism, and diseases change metabolic pathways leading to different metabolomic fingerprint ascribed to metabolites alterations [236, 244]. For instance, dietary habits as an external factor and gut microbiome as an internal factor have been demonstrated to be associated with susceptibility to development of some diseases like obesity and inflammatory bowel disease (IBD), respectively [245, 246].

Since the changes in bioenergetic metabolism such as glycolysis, tricarboxylic acid (TCA) cycle, and oxidative phosphorylation (OXPHOS) are one of the key features for tumor formation and progression, metabolomics provides a comprehensive toolbox for finding new diagnostic and therapeutic strategies in oncology area [247, 248]. Altogether it can be concluded that analysis of high-throughput data obtained from analytical and computational metabolomics approaches allows metabolic pathways annotation and the discovery of new molecules involved in biological and pathological process via systems [238].

4.2 Metabolomics Approaches

Following omics technologies evolution, metabolomics is thought as the most accurate and unbiased one due to precise measurement of metabolites involved in cellular processes. Metabolomics approaches try to measure physical properties of molecules such as their solubility in water or their polarity and then categorize them in subgroups of metabolites according to their functional groups or structural similarity [246, 249].

In order to unify the wide variety of cellular metabolites interpretation obtained from different analytical methods in different laboratories, standardization has been done according to some standard guidelines and some free databases like MetaboLights database that has developed by European scientists and Metabolomics Workbench that developed in USA [250–253].

Two key analytical approaches that are used in metabolomics are NMR spectroscopy and MS which could be coupled with either LC or GC intended for separation of metabolites of interest [254, 255]. Comprehensive data obtained from metabolomics methods need to be interpreted via chemometric principles for decoding and analyzing of results and discriminating metabolites present in each sample [256, 257].

4.2.1 Chromatographic-Based Methods

LC and GC which mainly used as separation systems for MS have their own pros and cons. In terms of resolution, GC is much more preferable, although the samples need to be volatized. Since a great deal of metabolites could not be vaporized, it is necessary to chemically modify molecules for volatizing. Despite GC, LC does not need derivatization that makes it remarkably more versatile technology than GC for separation of a broad range of small molecules [180, 236].

On the one hand in comparison with LC-MS and NMR, GC-MS needs the smallest sample volume, but on the other hand due to vaporization process in GC-MS the analytes are no longer recoverable. Additionally, the running and analysis time for complex mixture of samples via LC-MS is technically sophisticated [236]. The separation techniques like LC, GC, and capillary electrophoresis (CE) are very versatile approaches due to the multiple choices of stationary and mobile phases, as well as background electrolytes that are available. Due to mentioned advantages of LC-MS, it seems as the most common methods employing in metabolomic studies [218].

Among different types of LC, reversed-phase LC (RP-LC) with nonpolar stationary phase is suitable for hydrophobic molecules with medium to low polarity such as lipids, triglycerides, while normal-phase LC (NP-LC) and hydrophilic interaction LC (HILIC) with a polar stationary phase are preferable for separating polar compounds like amino acids, glycols, and sugars [218, 258, 259].

Different types of detectors like MS coupled with chromatographic strategies contribute to identification and classification of various metabolites with diverse properties [180]. The turning point of chromatographic methods was the advent of UHPLC that use much higher pressure for sample separation leading to faster running time, increased resolution, and the use of less solvent than other LC types [260, 261].

4.2.2 Capillary Electrophoresis (CE)

CE is another powerful analytical method for ions separation which could be coupled with MS instead of chromatographic approaches. As a further matter, in CE-MS the efficacy of metabolites peaks is drastically increased and thereby CE-MS sounds an extremely efficient, cost effective, and high-speed technique capable of separating a wide range of metabolites according to their electrophoretic mobility [218, 262].

Some overwhelming advantages of CE-MS over LC-MS are the capacity of CE-MS for rapid and uncomplicated separation process of analytes, its reproducibility and robustness, as well as feasibility of buffer compatibility for optimal ionization of sample [263].

4.2.3 Ion Mobility Spectrometry (IMS)

Chromatographic strategies and CE used as separation systems need a detector to visualize obtained data from sample analysis [240]. An emerging post-ionization separation approach is ion mobility spectrometry (IMS) which work via a gas phase to separate analytes according to their charge, size, and shape [264, 265]. When IMS is hyphenated with MS and LC called (LC-IM MS), boost the selectivity of separation process with no effect on the running time of ample. Thus, LC-IM MS is thought as a more precise method than simply LC-MS to identify the structural and conformational properties of given metabolites [266, 267].

4.2.4 Mass Spectrometry (MS)

The general framework of MS could be described as three main parts as follow, ion source, analyzer, and detector. The ions could be generated via different technologies in ion source that the most commonly used one includes matrix-assisted laser desorption/ionization (MALDI) and electrospray ionization (ESI) which both considered as soft ionization methods leading to minimum fragmentation of analytes and production of positive and negative ions. The ion fragments generated via ion source are then sorted in mass analyzer based on their mass-to-charge ratio (m/z). The last component of MS is responsible for the detection of electric signals produced by ion sorting. The final result of ion identification and quantification in MS is shown by a histogram indicating different m/z values of analyzed metabolites. MS is thought as a powerful tool with high selectivity which accurately analyzes metabolites present in a complex mixture based on their fragmentation patterns [268, 269].

MS-based approaches are the most widely used strategies in metabolomics for detecting the separated analytes based on m/z of charged ions. If a molecule does not have any charge, it must be ionized in an ion source in order to generate charged fragments able to be measured in MS [268, 270]. The combination of MS with LC, GC, or CE provides a versatile and powerful analytical tool for identification and quantification of metabolites in desired sample [240, 268, 270].

MS/MS or MS^2 is a more specified analytical technic composed of two or more types of analyzers for molecular weight determination and sequencing of peptides and other biomolecules in which ion precursors produced by first MS, ions are again fragmented in order to generate product ions via second MS for detection. Additional analyzer in MS/MS contributes to increased sample fragmentation by collision-induced dissociation (CID) that enhances the yield of ion fragments and finally improves the accuracy of diagrams which are indicator of metabolites of interest [271, 272].

4.2.5 Magnetic Resonance (NMR) Spectroscopy

NMR spectroscopy marked a turning point in structural clarification and identification of small molecules based on their proton content signals generated via nuclei activation in which magnetic field forces them to move out of alignment [268, 273]. This method made it possible to straightforwardly detect and quantify metabolites present in biological samples in a manner that is not destructive [274]. Moreover, due to the fact that NMR results are highly reproducible and the preparation and introduction of samples for this method are easy, NMR spectroscopy is thought as an appealing approach in diverse areas of metabolomics studies [275–277].

Despite immensely influential roles of NMR-based strategies in metabolite identification and quantification, it is considered as a moderately expensive technology with still low sensitivity [236, 268]. In order to improve NMR low sensitivity, the use of higher magnetic fields is suggested. It also need large amount of sample for each experiment that is regarded as another limiting factor when limited volume of sample is available [278, 279].

4.2.6 Fourier Transform Infrared Spectroscopy (FT-IR)

FT-IR is another method used in metabolomics area in which infrared absorbance spectrum of sample indicates the functional groups of metabolites [280]. The vibrational frequencies of analytes allow the fingerprinting of metabolome [281]. FT-IR is a precise method for profiling of biomolecules like amino acids, fatty acids, and carbohydrates, while its sensitivity and selectivity is lower than NMR and MS techniques [278].

4.2.7 Other Technologies

Imaging mass spectrometry (IMS), magnetic resonance spectroscopic imaging (MRSI), as well as magnetic resonance imaging (MRI) are capable to spatial mapping of lipids and peptides that allows uncovering functional properties of enzymes without destroying the cellular structure [217, 282]. Interestingly, high-resolution magic-angle spinning (HR-MAS) technology is suggested as an attractive strategy that supplies the concentration evaluation of metabolites in intact tissues [283, 284]. Furthermore, the hybrid MS/NMR could be a helpful method for identification of unknown analytes in untargeted studies [234].

Generally speaking, since each approach has its own advantages to measure a specific feature of metabolites or cellular process, the combination or coupling of various types of strategies is highly recommended to clarify identity, concentration, structural architecture, functional characteristics, and localization of small molecules.

4.3 Alteration in Metabolic Pathways of Stem Cells

Newly developed metabolomics approaches have open a new perspective in stem cell regulation and fate determination in terms of a variety of metabolites and biochemical reactions as key players of this process [285].

Glycolysis that occurs in cytosol and OXPHOS which takes place in mitochondria play crucial role in stem cell self-renewal, differentiation, and homeostasis. Due to the fact that stem cells are somehow similar to cancer cells, they are in need for rapid ATP production. To this end, glycolysis is an effective metabolic pathway composed of series of redox reactions take place in cytosol which is responsible for energy generation by means of glucose molecule breaks and its conversion into pyruvate or lactate molecules [286, 287].

The selective preference of using glycolysis as the main metabolic pathway for energy production designated Warburg effect. Warburg effect is a specific type of aerobic glycolysis and the main cause of high proliferation rate of stem cells leading to lactate production. By itself, glycolysis is considered as an anaerobic ATP generation metabolic pathway which occurs under either hypoxic condition or in absence of oxygen, whereas Warburg effect is attributed to selective glycolysis even in high oxygen concentration intended for maintenance of stemness property and rapid proliferation rate in stem cells and cancer cells [285, 288, 289].

Net energy production capacity of each glycolysis pathway is 2 ATP which comes from one six-carbon glucose molecule. Pyruvate is the final product of glycolysis and mainly could be converted to lactate or acetyl coenzyme A (acetyl-CoA) by lactate dehydrogenase (LDH) and pyruvate dehydrogenase (PDH) enzymes, respectively [285, 287]. The byproducts and intermediates produced followed by glycolysis could be used as building blocks of cellular macromolecules including nucleotides and proteins required for rapid proliferation of stem cells due to more rapid process of glycolysis than other energy producing metabolic pathways (Fig. 3) [289]. Here we focus on the main metabolic pathways involved in tuning different types of stem cell fate.

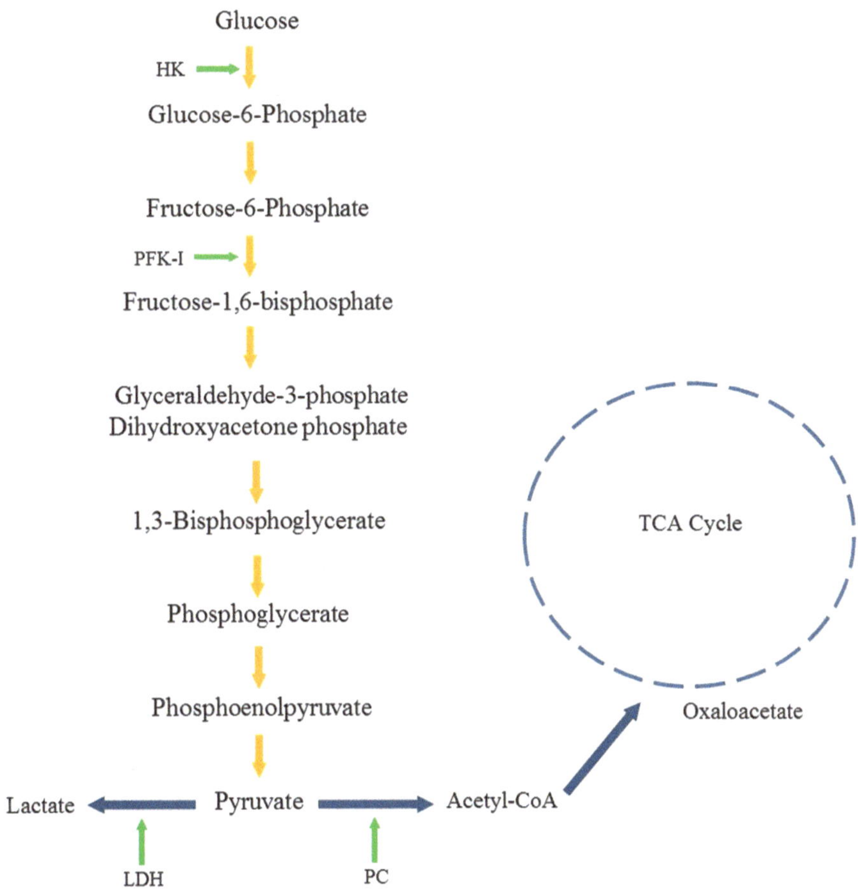

Fig. 3 Glycolysis is the most widely used pathway for energy production in cells like stem cells that undergo high proliferation rate. Although the net amount of ATP produced by glycolysis is only 2, much lower than TCA cycle, due to the fast growing nature of stem cells they prefer to use a rapid way for energy supplying that is glycolysis

As mentioned in previous sections, epigenetic regulation has a great importance in stem cell functions and stem cell metabolite pool is able to regulate epigenetic factors. In pluripotent stem cells some metabolites affect the methylation and acetylation state of chromatin. In terms of lipids, it has been reported that the concentration of unsaturated fatty acids in pluripotent stem cells is high and eicosanoid synthesis pathways are suppressed. Moreover, during differentiation process concomitant with increased OXPHOS activity and decreased glycolysis, oxidation of unsaturated fatty acids leads to drop in unsaturated metabolites abundance [286].

4.3.1 Totipotent Stem Cells

In totipotent stem cells which referred to early blastomere before morula stage, due to low activity of two essential rate-limiting enzymes of glycolysis, hexokinase (HK) and phosphofructokinase-1 (PFK-1), the initial rate of glycolysis is not high and pyruvate considered as the key energy generation source [287]. In order to either anaplerosis or gluconeogenesis pyruvate converts to acetyl-CoA and oxaloacetate using PDH and pyruvate carboxylase (PC), respectively. TCA cycle occurs in the matrix of mitochondria and produces different intermediates of carbon resulting in electron transport chain (ETC) activation in inner membrane of mitochondria. It is worthy of note that although totipotent stem cells exhibit OXPHOS, they show lower oxygen consumption rate than differentiated cells [287, 290].

Due to lack of spherical mitochondria replication in early embryo, rapid proliferation of totipotent stem cells leads to a gradual decrease in ATP synthesis. Bicarbonate and other ions transport to totipotent stem cells and contribute to oxaloacetate synthesis via PC needed for anaplerosis and providing nucleotides for production of DNA and RNA using carbamoyl-phosphate synthetase 2, aspartate transcarbamoylase, and dihydroorotase (CAD complex). Therefore, pyruvate oxidation and bicarbonate fixation sound as the main metabolic pathways in totipotent stem cells [287].

4.3.2 Pluripotent Stem Cells

In pluripotent stem cells like ESCs, increased flux of glycolysis reactions is the consequence of higher glucose transporter type 1 (GLUT1) and GLUT3 expression levels which lead to enhanced glucose uptake by these cells [291, 292]. Moreover, two essential rate-limiting enzymes of glycolysis including HK and PFK-1 exhibit high activity that highlights the priority of this metabolic pathway in pluripotent stem cells leading to promoted nucleotide synthesis via pentose phosphate pathway (PPP) required for rapid nucleotide production for high rate proliferation of pluripotent stem cells [291, 293].

Owing to the Warburg effect, glycolysis is supposed to provide the building block for biosynthesis of essential molecules including nucleic acids, proteins, and lipids needed for rapid proliferation [289]. In pluripotent stem cells, glycolysis pathway is the predominant source for ATP production despite its lower efficacy in comparison with TCA. Even during reprograming of somatic cells, the carbon source for ATP generation shift from OXPHOS in differentiated cells to glycolysis in iPSCs [294, 295]. It has been reported that iPSC reprogramming factor will be upregulated following increased level of glycolysis [294, 296].

LIN28A and LIN28B, two kinds of RNA-binding proteins, are known as the key factors for stemness maintenance in ESCs, fetal HSCs, neural crest progenitors, spermatogonia, and skeletal myoblasts during muscle regeneration. They are also essential elements which function as a reprograming factor through iPSCs production which are capable to bind to mRNA molecules of a variety of glycolytic enzymes and mitochondrial genes [297–299].

It has been demonstrated that LIN28A bind to tumor-suppressive let-7 pre-miRNA inhibits the let-7 microRNA maturations leading to pluripotency maintenance by regulating the glucose metabolism [300, 301]. These reports indicate the crucial roles of LIN28A and LIN28B in mammalian stemness regulation through different lineages [287, 291].

In ESCs mitochondrial oxidation is essential for nicotinamide adenine dinucleotide (NAD)+ recycle in order to replenish TCA cycle for fatty acids and amino acids production [294, 302]. PPP also shows increased level in ESCs as a supplier for ribose 5 phosphate and nicotinamide adenine dinucleotide phosphate (NADPH) required for rapid proliferation of DNA, RNA, as well as fatty acids synthesis [293, 303].

Acetyl-CoA contributes to acetylate histone H3 that is considered as an epigenetic code for active euchromatin state responsible for keeping the pluripotent stem cells in pluripotent state either in ESCs or in iPSCs [304, 305]. Acetyl-CoA produced via active glycolytic reactions in cytosol and acetate, as a cytosolic acetyl-CoA precursor, seems to be an inhibitor of histone deacetylation of histone H3 and consequently as a suppressor of pluripotent stem cells differentiation [285]. Therefore, glycolysis and also its derivatives and intermediates are involved in self-renewal regulation of pluripotent stem cells [306].

4.3.3 Multipotent Stem Cells

In multipotent stem cells like HSCs and MSCs that locate in a hypoxic niche, the lower concentration of oxygen can assist their stemness state and regulate cell fate [307, 308]. Increased levels of glycolytic enzymes indicate an adaption of HSCs and MSCs to hypoxic environment in the bone marrow with lower oxygen saturation than venous blood [308, 309].

Hypoxia-inducible factor-1α (HIF1α) acts as a regulator of transcription which adapts cells with hypoxic conditions by involvement in cellular metabolic pathways. HIF1α upregulates under hypoxic conditions and its high expression level has been reported in HSCs that reside in hypoxic niche [291, 310, 311].

Reactive oxygen species (ROS) are harmful reactive by-products of oxidative metabolism and are thought to impose tissue damage and are shown to promote cell differentiation. Adipose-derived stem cells which are categorized as a type of MSCs try to prevent ROS-induced cellular damage and consequently maintain the stemness potential of adipose-derived stem cells [308, 312]. Similar to other types of stem cells, adipose-derived stem cells show upregulation of glycolytic enzymes explaining the great importance of glycolysis for maintaining stemness state in comparison with OXPHOS even in quiescent state [291].

MSCs culture under normoxia leads to promoted proliferation rate and approximately 4 times increased senescence that comes from shifting glycolysis to OXPHOS. Therefore, it sounds that glycolysis under hypoxic niche ensures long-term stemness maintenance of MSCs and suppresses their ROS-induced differentiation [313].

Adult stem cells as multipotent stem cells such as long-term hematopoietic stem cells (LT-HSCs), MSCs, and neural stem cells (NSCs) residing in the quiescent state exhibit pivotal role in damaged tissue regeneration and renewal. Contrary to ESCs, it has been revealed that adult stem cells show slow-cycling paralleled with self-renewal properties essential for their regenerative functions and prevention from ROS-induced cellular damage [308, 312]. In LT-HSCs functionally immature mitochondria cause lower levels of ROS that seems to be relevant to increased levels of glycolytic enzymes stimulated by hypoxic environment similar to cancer cells [287, 314]. It has been demonstrated that high levels of ROS production in LT-HSCs and NSCs in hypoxic niche impose impaired self-renewal and lead to their differentiation or apoptosis [315, 316].

This glycolytic induced phenotype in LT-HSCs is attributed to the function of HSC transcription factor called homeobox protein MEIS1 by HIF1α, as its target, which controls the upregulation of numbers of enzymes associated with glycolysis [317]. In addition, it has been reported that fatty acid β-oxidation is essential for maintenance of self-renewal and differentiation inhibition of LT-HSCs, NSCs, and skeletal muscle stem cells (SMSCs), thereby different types of fatty acids are catabolized in order to produce acetyl-CoA, NADH, and flavin adenine dinucleotide (FADH2) for fueling OXPHOS via promoting TCA cycle. The oxidation of NADH and FADH2 by four protein complexes of ETC function in mitochondria is the source of energy generated from fatty acid β-oxidation [318, 319]. Fukawa et al. reveal that excessive fatty acid β-oxidation adversely affects the SMSCs that led to muscle wasting and atrophy indicating that optimal level of fatty acid β-oxidation is needed for multipotency capacity maintenance of SMSCs [320].

Owing to immature mitochondrial inner membrane, caused by impaired OXPHOS, pyruvate dehydrogenase kinase 1 (PDK1) and PDK3 are upregulated in LT-HSCs which suppress PDH, thereby acetyl-CoA production from lactate via PDH is inhibited. PDK prevents pyruvate oxidation and therefore TCA cycle is impaired and lactate is considered as the final product of glycolysis since reduced production of acetyl-CoA by means of PDH action [321]. Other isoforms of PDK named PDK2 and PDK4 are critical factors for self-renewal property of LT-HSCs since they are regulated via HIF1α. These reports revealed that OXPHOS is decreased LT-HSCs and the predominant metabolic pathway in LT-HSCs is glycolysis leading to the generation of lactate from pyruvate instead of pyruvate shunt to TCA cycle (Fig. 4) [322].

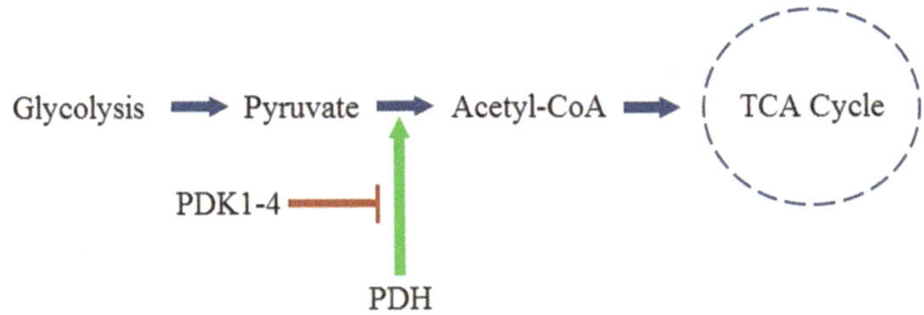

Fig. 4 Impaired OXPHOS in LT-HSCs caused by inhibitory effect of isoforms of PDK. PDK 1–4 suppress the PDH activity, thereby pyruvate could not produce acetyl-CoA. Therefore, TCA cycle and subsequent OXPHOS are reduced

According to unequivocal evidence for metabolomics-based studies on adult stem cells under hypoxic environment, it has been suggested that among different quiescent stem cell types in LT-HSCs, MSCs, and NSCs, OXPHOS and ROS production in normoxia could be a trigger for loss of self-renewal potential and senescence induction [291, 316, 323].

Taken together, breakthroughs in metabolomics approaches including GC-MS, LC-MS, and NMR imaging are thought as the most reliable analytical methods intended for metabolites measurement for uncovering different metabolic pathways involved in stem cell regulating processes.

5 Conclusion and Future Perspectives

Recent achievements have demonstrated that any trait of each cell arises from genomic material that in turn is the consequence of its expression to proteins and ultimately the final product of proteins function that are metabolites. NGS as a developed sequencing method allows the fast whole-genome sequencing of various types of cells and organisms which boost our knowledge of cellular genetic instructions contributing the interpretation of cellular behavior according to their genetic content. Genome editing methods, CRISPR/Cas9 in particular, in combination with iPSCs biology progression try to engineer patient-specific disease models and offer new therapeutic approaches by means of comparative studies. Moreover, these methods make it possible to manipulate the genome intended for its customized engineering in order to correct the mutant sequences present in each disease.

Post-genomic studies such as transcriptomics, proteomics, and metabolomics seem more sophisticated science branches than simply genomics that is due to their dynamic nature and immense diversity. Among these technologies that elucidate and decode the RNA, proteins, and metabolites structures and functions, metabolomics considered as the most explicit one for interpreting cellular processes. In terms of stem cells, the abundance and concentrations of each analyte in the cell context is the key player that precisely regulates cell response to internal and environmental stimuli and controls their self-renewal or differentiation state. Therefore, thanks to omics strategies implementation we are capable of manipulating stem cells for our targeted studies.

Metabolomics, the latest-omics technology, provides detailed instructions to reveal exact biochemical reactions and analytes governing the phenotype of different types of stem cells such as ESCs, adult stem cells, and iPSCs. This evidence results in discovering metabolic pathways alterations responsible for molecular events of each cell type. Furthermore, MS- and NMR-based technologies as computational and analytical methods measuring the presence and quantity of metabolites that enable us to introduce specific biomarkers attributed to stemness state maintenance or differentiation trigger drug and pathological agents response.

Owing to the importance of metabolites in cell fate determination, high-throughput metabolite profiling of various stem cell types provides unprecedented level of molecular resolution which is hoped with the aid of bioinformatics and biological databases predict biochemical pathways essential for regulation of cellular structure and functions. This makes it possible to predict cell behavior based on multi-purpose strategies and produce more efficient and targeted therapies according to personalized medicine. Interestingly, biomarkers sound as prognostic approaches that in combination with other omics data, remarkably reduce the burden of diseases.

Taken together, multi-omics disciplines incorporated with stem cell developments are thought as powerful tools for disclosing fundamental molecular pathways governing cellular process which lead to providing novel approaches for personalizing the prevention, prognosis, diagnosis, and efficient targeted therapies of incurable diseases specific for each individual.

References

1. Winkler J, Sotiriadou I, Chen S, Hescheler J, Sachinidis A. The potential of embryonic stem cells combined with -omics technologies as model systems for toxicology. Curr Med Chem. 2009;16(36):4814–27.
2. Arrell DK, Terzic A. Systems proteomics for translational network medicine. Circ Cardiovasc Genet. 2012;5(4):478.
3. Faustino RS, Arrell DK, Folmes CD, Terzic A, Perez-Terzic C. Stem cell systems informatics for advanced clinical biodiagnostics: tracing molecular signatures from bench to bedside. Croat Med J. 2013;54(4):319–29.
4. Wyles SP, Faustino RS, Li X, Terzic A, Nelson TJ. Systems-based technologies in profiling the stem cell molecular framework for cardioregenerative medicine. Stem Cell Rev Rep. 2015;11(3):501–10.
5. Gerrits A, Dykstra B, Otten M, Bystrykh L, de Haan G. Combining transcriptional profiling and genetic linkage analysis to uncover gene networks operating in hematopoietic stem cells and their progeny. Immunogenetics. 2008;60(8):411–22.
6. De Los Angeles A, Ferrari F, Xi R, Fujiwara Y, Benvenisty N, Deng H, Hochedlinger K, Jaenisch R, Lee S, Leitch HG. Hallmarks of pluripotency. Nature. 2015;525(7570):469.
7. Hima Bindu A, Srilatha B. Potency of various types of stem cells and their transplantation. J Stem Cell Res Ther. 2011;1:115.

8. Hanna J, Cheng AW, Saha K, Kim J, Lengner CJ, Soldner F, Cassady JP, Muffat J, Carey BW, Jaenisch R. Human embryonic stem cells with biological and epigenetic characteristics similar to those of mouse ESCs. Proc Natl Acad Sci. 2010;107(20):9222–7.
9. Khanlarkhani N, Baazm M, Mohammadzadeh F, Najafi A, Mehdinejadiani S, Sobhani A. Multipotent stem cell and reproduction. J Stem Cells. 2016;11(4):219.
10. Sobhani A, Khanlarkhani N, Baazm M, Mohammadzadeh F, Najafi A, Mehdinejadiani S, Aval FS. Multipotent stem cell and current application. Acta Med Iran. 2017;55(1):6–23.
11. de Kretser D. Totipotent, pluripotent or unipotent stem cells: a complex regulatory enigma and fascinating biology. J Law Med. 2007;15(2):212.
12. Goodarzi P, Falahzadeh K, Aghayan H, Payab M, Larijani B, Alavi-Moghadam S, Tayanloo-Beik A, Adibi H, Gilany K, Arjmand B. Therapeutic abortion and ectopic pregnancy: alternative sources for fetal stem cell research and therapy in Iran as an Islamic country. Cell Tissue Bank. 2019;20(1):11–24.
13. Bai Q, Desprat R, Klein B, Lemaitre J-M, De Vos J. Embryonic stem cells or induced pluripotent stem cells? A DNA integrity perspective. Curr Gene Ther. 2013;13(2):93–8.
14. Kalra K, Tomar P. Stem cell: basics, classification and applications. Am J Phytomed Clin Ther. 2014;2(7):919–30.
15. Zhan M. Genomic studies to explore self-renewal and differentiation properties of embryonic stem cells. Front Biosci. 2008;13(13):276–83.
16. O'Donoghue K, Fisk NM. Fetal stem cells. Best Pract Res Clin Obstet Gynaecol. 2004;18(6):853–75.
17. Pappa KI, Anagnou NP. Novel sources of fetal stem cells: where do they fit on the developmental continuum? Regen Med. 2009;4(3):423–33.
18. NIH. Stem cells: scientific progress and future research directions. In: Stem cell information. Bethesda: National Institutes of Health, U.S. Department of Health and Human Services; 2001.
19. Zhao Y, Glesne D, Huberman E. A human peripheral blood monocyte-derived subset acts as pluripotent stem cells. Proc Natl Acad Sci. 2003;100(5):2426–31.
20. Chagastelles PC, Nardi NB. Biology of stem cells: an overview. Kidney Int Suppl. 2011;1(3):63–7.
21. Crisan M, Yap S, Casteilla L, Chen C-W, Corselli M, Park TS, Andriolo G, Sun B, Zheng B, Zhang L. A perivascular origin for mesenchymal stem cells in multiple human organs. Cell Stem Cell. 2008;3(3):301–13.
22. Ullah I, Subbarao RB, Rho GJ. Human mesenchymal stem cells-current trends and future prospective. Biosci Rep. 2015;35(2):e00191.
23. Lanza R, Langer R, Vacanti JP. Principles of tissue engineering. Cambridge: Academic press; 2011.
24. Soria B, Pezzolla D, López J, Rojas A, Hmadcha A. Generation of pancreatic islets from stem cells. In: Principles of tissue engineering. 4th ed. Cambridge: Elsevier; 2014. p. 837–47.
25. Loh Y-H, Yang L, Yang JC, Li H, Collins JJ, Daley GQ. Genomic approaches to deconstruct pluripotency. Annu Rev Genomics Hum Genet. 2011;12:165–85.
26. Hasin Y, Seldin M, Lusis A. Multi-omics approaches to disease. Genome Biol. 2017;18(1):83.
27. Bertin PN, Médigue C, Normand P. Advances in environmental genomics: towards an integrated view of micro-organisms and ecosystems. Microbiology. 2008;154(2):347–59.
28. Fleischmann RD, Adams MD, White O, Clayton RA, Kirkness EF, Kerlavage AR, Bult CJ, Tomb J-F, Dougherty BA, Merrick JM. Whole-genome random sequencing and assembly of Haemophilus influenzae Rd. Science. 1995;269(5223):496–512.
29. International Human Genome Sequencing Consortium. Initial sequencing and analysis of the human genome. Nature. 2001;409(6822):860.
30. Hamazaki T, El Rouby N, Fredette NC, Santostefano KE, Terada N. Concise review: induced pluripotent stem cell research in the era of precision medicine. Stem Cells. 2017;35(3):545–50.
31. Venter JC, Adams MD, Myers EW, Li PW, Mural RJ, Sutton GG, Smith HO, Yandell M, Evans CA, Holt RA. The sequence of the human genome. Science. 2001;291(5507):1304–51.

32. Watson JD. The human genome project: past, present, and future. Science. 1990;248(4951):44–9.
33. International HapMap Consortium. A second generation human haplotype map of over 3.1 million SNPs. Nature. 2007;449(7164):851.
34. Hapmap C. The international HapMap project: the international HapMap consortium. Nature. 2003;426:789–96.
35. 1000 Genomes Project Consortium. A global reference for human genetic variation. Nature. 2015;526(7571):68.
36. Voight BF, Kang HM, Ding J, Palmer CD, Sidore C, Chines PS, Burtt NP, Fuchsberger C, Li Y, Erdmann J. The metabochip, a custom genotyping array for genetic studies of metabolic, cardiovascular, and anthropometric traits. PLoS Genet. 2012;8(8):e1002793.
37. Wang WY, Barratt BJ, Clayton DG, Todd JA. Genome-wide association studies: theoretical and practical concerns. Nat Rev Genet. 2005;6(2):109.
38. Cirulli ET, Goldstein DB. Uncovering the roles of rare variants in common disease through whole-genome sequencing. Nat Rev Genet. 2010;11(6):415.
39. Hockemeyer D, Soldner F, Beard C, Gao Q, Mitalipova M, DeKelver RC, Katibah GE, Amora R, Boydston EA, Zeitler B. Efficient targeting of expressed and silent genes in human ESCs and iPSCs using zinc-finger nucleases. Nat Biotechnol. 2009;27(9):851.
40. Koboldt DC, Steinberg KM, Larson DE, Wilson RK, Mardis ER. The next-generation sequencing revolution and its impact on genomics. Cell. 2013;155(1):27–38.
41. Gasperskaja E, Kučinskas V. The most common technologies and tools for functional genome analysis. Acta Med Litu. 2017;24(1):1.
42. Bickel PJ, Brown JB, Huang H, Li Q. An overview of recent developments in genomics and associated statistical methods. Philos Trans R Soc A Math Phys Eng Sci. 2009;367(1906):4313–37.
43. McCreath SB, Delgoda R. Pharmacognosy: fundamentals, applications and strategies. Cambridge: Academic Press; 2017.
44. Brown T. Southern blotting. Curr Protoc Protein Sci. 1998;13(1):A.4G.
45. Mahmood T, Yang P-C. Western blot: technique, theory, and trouble shooting. N Am J Med Sci. 2012;4(9):429–34.
46. Van Oss C, Good R, Chaudhury M. Mechanism of DNA (southern) and protein (Western) blotting on cellulose nitrate and other membranes. J Chromatogr A. 1987;391:53–65.
47. Garibyan L, Avashia N. Polymerase chain reaction. J Invest Dermatol. 2013;133(3):1–4.
48. Singh J, Birbian N, Sinha S, Goswami A. A critical review on PCR, its types and applications. Int J Adv Res Biol Sci. 2014;1(7):65–80.
49. Clark DP, Pazdernik NJ. Molecular biology. Amsterdam: Elsevier; 2012.
50. Wanger A, Chavez V, Huang R, Wahed A, Dasgupta A, Actor JK. Microbiology and molecular diagnosis in pathology: a comprehensive review for board preparation, certification and clinical practice. Amsterdam: Elsevier; 2017.
51. Schena M, Shalon D, Davis RW, Brown PO. Quantitative monitoring of gene expression patterns with a complementary DNA microarray. Science. 1995;270(5235):467–70.
52. Bulyk ML, Johnson PL, Church GM. Nucleotides of transcription factor binding sites exert interdependent effects on the binding affinities of transcription factors. Nucleic Acids Res. 2002;30(5):1255–61.
53. Sabo PJ, Kuehn MS, Thurman R, Johnson BE, Johnson EM, Cao H, Yu M, Rosenzweig E, Goldy J, Haydock A. Genome-scale mapping of DNase I sensitivity in vivo using tiling DNA microarrays. Nat Methods. 2006;3(7):511.
54. Hagemann IS, Devarakonda S, Lockwood CM, Spencer DH, Guebert K, Bredemeyer AJ, Al-Kateb H, Nguyen TT, Duncavage EJ, Cottrell CE. Clinical next-generation sequencing in patients with non–small cell lung cancer. Cancer. 2015;121(4):631–9.
55. Sanger F, Air GM, Barrell BG, Brown NL, Coulson AR, Fiddes JC, Hutchison CA III, Slocombe PM, Smith M. Nucleotide sequence of bacteriophage φX174 DNA. Nature. 1977;265(5596):687.

56. Heather JM, Chain B. The sequence of sequencers: the history of sequencing DNA. Genomics. 2016;107(1):1–8.
57. Hall N. Advanced sequencing technologies and their wider impact in microbiology. J Exp Biol. 2007;210(9):1518–25.
58. Ke R, Mignardi M, Hauling T, Nilsson M. Fourth generation of next-generation sequencing technologies: promise and consequences. Hum Mutat. 2016;37(12):1363–7.
59. Carey MF, Peterson CL, Smale ST. Chromatin immunoprecipitation (chip). Cold Spring Harb Protoc. 2009;2009(9):pdb.prot5279.
60. Gade P, Kalvakolanu DV. Chromatin immunoprecipitation assay as a tool for analyzing transcription factor activity. In: Transcriptional regulation. New York: Springer; 2012. p. 85–104.
61. Dekker J, Rippe K, Dekker M, Kleckner N. Capturing chromosome conformation. Science. 2002;295(5558):1306–11.
62. Ho JW, Bishop E, Karchenko PV, Nègre N, White KP, Park PJ. ChIP-chip versus ChIP-seq: lessons for experimental design and data analysis. BMC Genomics. 2011;12(1):134.
63. McDaniell R, Lee B-K, Song L, Liu Z, Boyle AP, Erdos MR, Scott LJ, Morken MA, Kucera KS, Battenhouse A. Heritable individual-specific and allele-specific chromatin signatures in humans. Science. 2010;328(5975):235–9.
64. Grskovic M, Javaherian A, Strulovici B, Daley GQ. Induced pluripotent stem cells—opportunities for disease modelling and drug discovery. Nat Rev Drug Discov. 2011;10(12):915.
65. Takahashi K, Tanabe K, Ohnuki M, Narita M, Ichisaka T, Tomoda K, Yamanaka S. Induction of pluripotent stem cells from adult human fibroblasts by defined factors. Cell. 2007;131(5):861–72.
66. Bock C, Kiskinis E, Verstappen G, Gu H, Boulting G, Smith ZD, Ziller M, Croft GF, Amoroso MW, Oakley DH. Reference maps of human ES and iPS cell variation enable high-throughput characterization of pluripotent cell lines. Cell. 2011;144(3):439–52.
67. Hendriks WT, Warren CR, Cowan CA. Genome editing in human pluripotent stem cells: approaches, pitfalls, and solutions. Cell Stem Cell. 2016;18(1):53–65.
68. Maeder ML, Gersbach CA. Genome-editing technologies for gene and cell therapy. Mol Ther. 2016;24(3):430–46.
69. Zhang Z, Zhang Y, Gao F, Han S, Cheah KS, Tse H-F, Lian Q. CRISPR/Cas9 genome-editing system in human stem cells: current status and future prospects. Mol Ther Nucleic Acids. 2017b;9:230–41.
70. Joung JK, Sander JD. TALENs: a widely applicable technology for targeted genome editing. Nat Rev Mol Cell Biol. 2013;14(1):49.
71. Mali P, Esvelt KM, Church GM. Cas9 as a versatile tool for engineering biology. Nat Methods. 2013;10(10):957.
72. Urnov FD, Rebar EJ, Holmes MC, Zhang HS, Gregory PD. Genome editing with engineered zinc finger nucleases. Nat Rev Genet. 2010;11(9):636.
73. Zou J, Maeder ML, Mali P, Pruett-Miller SM, Thibodeau-Beganny S, Chou B-K, Chen G, Ye Z, Park I-H, Daley GQ. Gene targeting of a disease-related gene in human induced pluripotent stem and embryonic stem cells. Cell Stem Cell. 2009;5(1):97–110.
74. Boch J, Scholze H, Schornack S, Landgraf A, Hahn S, Kay S, Lahaye T, Nickstadt A, Bonas U. Breaking the code of DNA binding specificity of TAL-type III effectors. Science. 2009;326(5959):1509–12.
75. Ding Q, Lee Y-K, Schaefer EA, Peters DT, Veres A, Kim K, Kuperwasser N, Motola DL, Meissner TB, Hendriks WT. A TALEN genome-editing system for generating human stem cell-based disease models. Cell Stem Cell. 2013;12(2):238–51.
76. Luo Y, Rao M, Zou J. Generation of GFP reporter human induced pluripotent stem cells using aavs1 safe harbor transcription activator-like effector nuclease. Curr Protoc Stem Cell Biol. 2014;29(1):5A.7.1–5A.7.18.
77. Soldner F, Laganière J, Cheng AW, Hockemeyer D, Gao Q, Alagappan R, Khurana V, Golbe LI, Myers RH, Lindquist S. Generation of isogenic pluripotent stem cells differing exclusively at two early onset Parkinson point mutations. Cell. 2011;146(2):318–31.

78. Thurtle-Schmidt DM, Lo TW. Molecular biology at the cutting edge: a review on CRISPR/CAS9 gene editing for undergraduates. Biochem Mol Biol Educ. 2018;46(2):195–205.
79. Mojica FJ, Díez-Villaseñor C, Soria E, Juez G. Biological significance of a family of regularly spaced repeats in the genomes of Archaea, Bacteria and mitochondria. Mol Microbiol. 2000;36(1):244–6.
80. Van Der Oost J, Westra ER, Jackson RN, Wiedenheft B. Unravelling the structural and mechanistic basis of CRISPR–Cas systems. Nat Rev Microbiol. 2014;12(7):479.
81. Wang H, La Russa M, Qi LS. CRISPR/Cas9 in genome editing and beyond. Annu Rev Biochem. 2016;85:227–64.
82. Andrey G, Spielmann M. CRISPR/Cas9 genome editing in embryonic stem cells. In: Enhancer RNAs. New York: Springer; 2017. p. 221–34.
83. Marraffini LA. CRISPR-Cas immunity in prokaryotes. Nature. 2015;526(7571):55.
84. Nishimasu H, Ran FA, Hsu PD, Konermann S, Shehata SI, Dohmae N, Ishitani R, Zhang F, Nureki O. Crystal structure of Cas9 in complex with guide RNA and target DNA. Cell. 2014;156(5):935–49.
85. Kunin V, Sorek R, Hugenholtz P. Evolutionary conservation of sequence and secondary structures in CRISPR repeats. Genome Biol. 2007;8(4):R61.
86. Makarova KS, Haft DH, Barrangou R, Brouns SJ, Charpentier E, Horvath P, Moineau S, Mojica FJ, Wolf YI, Yakunin AF. Evolution and classification of the CRISPR–Cas systems. Nat Rev Microbiol. 2011;9(6):467.
87. Musunuru K. Genome editing of human pluripotent stem cells to generate human cellular disease models. Dis Model Mech. 2013;6(4):896–904.
88. Dimos JT, Rodolfa KT, Niakan KK, Weisenthal LM, Mitsumoto H, Chung W, Croft GF, Saphier G, Leibel R, Goland R. Induced pluripotent stem cells generated from patients with ALS can be differentiated into motor neurons. Science. 2008;321(5893):1218–21.
89. Seki T, Yuasa S, Oda M, Egashira T, Yae K, Kusumoto D, Nakata H, Tohyama S, Hashimoto H, Kodaira M. Generation of induced pluripotent stem cells from human terminally differentiated circulating T cells. Cell Stem Cell. 2010;7(1):11–4.
90. Liu G-H, Qu J, Suzuki K, Nivet E, Li M, Montserrat N, Yi F, Xu X, Ruiz S, Zhang W. Progressive degeneration of human neural stem cells caused by pathogenic LRRK2. Nature. 2012;491(7425):603.
91. Reinhardt P, Schmid B, Burbulla LF, Schöndorf DC, Wagner L, Glatza M, Höing S, Hargus G, Heck SA, Dhingra A. Genetic correction of a LRRK2 mutation in human iPSCs links parkinsonian neurodegeneration to ERK-dependent changes in gene expression. Cell Stem Cell. 2013;12(3):354–67.
92. Deleidi M, Yu C. Genome editing in pluripotent stem cells: research and therapeutic applications. Biochem Biophys Res Commun. 2016;473(3):665–74.
93. Tebas P, Stein D, Tang WW, Frank I, Wang SQ, Lee G, Spratt SK, Surosky RT, Giedlin MA, Nichol G. Gene editing of CCR5 in autologous CD4 T cells of persons infected with HIV. N Engl J Med. 2014;370(10):901–10.
94. Chang C-W, Lai Y-S, Westin E, Khodadadi-Jamayran A, Pawlik KM, Lamb LS Jr, Goldman FD, Townes TM. Modeling human severe combined immunodeficiency and correction by CRISPR/Cas9-enhanced gene targeting. Cell Rep. 2015;12(10):1668–77.
95. Cong L, Ran FA, Cox D, Lin S, Barretto R, Habib N, Hsu PD, Wu X, Jiang W, Marraffini L. Multiplex genome engineering using CRISPR/Cas systems. Science. 2013;339(6121):819–23.
96. Boj SF, Hwang C-I, Baker LA, Chio IIC, Engle DD, Corbo V, Jager M, Ponz-Sarvise M, Tiriac H, Spector MS. Organoid models of human and mouse ductal pancreatic cancer. Cell. 2015;160(1):324–38.
97. Huch M, Boj SF, Clevers H. Lgr5+ liver stem cells, hepatic organoids and regenerative medicine. Regen Med. 2013a;8(4):385–7.
98. Huch M, Koo B-K. Modeling mouse and human development using organoid cultures. Development. 2015;142(18):3113–25.

99. Drost J, Clevers H. Translational applications of adult stem cell-derived organoids. Development. 2017;144(6):968–75.
100. Huch M, Bonfanti P, Boj SF, Sato T, Loomans CJ, Van De Wetering M, Sojoodi M, Li VS, Schuijers J, Gracanin A. Unlimited in vitro expansion of adult bi-potent pancreas progenitors through the Lgr5/R-spondin axis. EMBO J. 2013b;32(20):2708–21.
101. Takata N, Eiraku M. Stem cells and genome editing: approaches to tissue regeneration and regenerative medicine. J Hum Genet. 2018;63(2):165.
102. Deans C, Maggert KA. What do you mean,"epigenetic"? Genetics. 2015;199(4):887–96.
103. Weinhold B. Epigenetics: the science of change. Environ Health Perspect. 2006;114(3):A160–7.
104. Chung H, Sidhu KS. Epigenetic modifications of embryonic stem cells: current trends and relevance in developing regenerative medicine. Stem Cells Cloning. 2008;1:11.
105. Efroni S, Duttagupta R, Cheng J, Dehghani H, Hoeppner DJ, Dash C, Bazett-Jones DP, Le Grice S, McKay RD, Buetow KH. Global transcription in pluripotent embryonic stem cells. Cell Stem Cell. 2008;2(5):437–47.
106. Kraushaar DC, Zhao K. The epigenomics of embryonic stem cell differentiation. Int J Biol Sci. 2013;9(10):1134.
107. Kraushaar DC, Dalton S, Wang L. Heparan sulfate: a key regulator of embryonic stem cell fate. Biol Chem. 2013;394(6):741–51.
108. Lanner F, Rossant J. The role of FGF/Erk signaling in pluripotent cells. Development. 2010;137(20):3351–60.
109. Lunyak VV, Rosenfeld MG. Epigenetic regulation of stem cell fate. Hum Mol Genet. 2008;17(R1):R28–36.
110. Rodriguez RT, Velkey JM, Lutzko C, Seerke R, Kohn DB, O'Shea KS, Firpo MT. Manipulation of OCT4 levels in human embryonic stem cells results in induction of differential cell types. Exp Biol Med. 2007;232(10):1368–80.
111. Zaehres H, Schöler HR. Induction of pluripotency: from mouse to human. Cell. 2007;131(5):834–5.
112. Piunti A, Shilatifard A. Epigenetic balance of gene expression by Polycomb and COMPASS families. Science. 2016;352(6290):aad9780.
113. Parsons XH. Embedding the future of regenerative medicine into the open epigenomic landscape of pluripotent human embryonic stem cells. Ann Res Rev Biol. 2013;3(4):323.
114. Raisner RM, Hartley PD, Meneghini MD, Bao MZ, Liu CL, Schreiber SL, Rando OJ, Madhani HD. Histone variant H2A. Z marks the 5′ ends of both active and inactive genes in euchromatin. Cell. 2005;123(2):233–48.
115. Reik W, Dean W, Walter J. Epigenetic reprogramming in mammalian development. Science. 2001;293(5532):1089–93.
116. Zhou Y, Kim J, Yuan X, Braun T. Epigenetic modifications of stem cells: a paradigm for the control of cardiac progenitor cells. Circ Res. 2011;109(9):1067–81.
117. Bird A. DNA methylation patterns and epigenetic memory. Genes Dev. 2002;16(1):6–21.
118. Zhu Q, Wani AA. Histone modifications: crucial elements for damage response and chromatin restoration. J Cell Physiol. 2010;223(2):283–8.
119. Zhao Y, Garcia BA. Comprehensive catalog of currently documented histone modifications. Cold Spring Harb Perspect Biol. 2015;7(9):a025064.
120. Parsons XH, Garcia SN, Pillus L, Kadonaga JT. Histone deacetylation by Sir2 generates a transcriptionally repressed nucleoprotein complex. Proc Natl Acad Sci. 2003;100(4):1609–14.
121. Sterner DE, Berger SL. Acetylation of histones and transcription-related factors. Microbiol Mol Biol Rev. 2000;64(2):435–59.
122. Haushalter KA, Kadonaga JT. Chromatin assembly by DNA-translocating motors. Nat Rev Mol Cell Biol. 2003;4(8):613.
123. Fyodorov DV, Kadonaga JT. The many faces of chromatin remodeling: SWItching beyond transcription. Cell. 2001;106(5):523–5.
124. Mazina MY, Vorobyeva N. The role of ATP-dependent chromatin remodeling complexes in regulation of genetic processes. Russ J Genet. 2016;52(5):463–72.

125. Wary A, Wary N, Baruah J, Mastej V, Wary KK. Chromatin-modifying agents convert fibroblasts to OCT4+ and VEGFR-2+ capillary tube-forming cells. PLoS One. 2017;12(5):e0176496.
126. Azuara V, Perry P, Sauer S, Spivakov M, Jørgensen HF, John RM, Gouti M, Casanova M, Warnes G, Merkenschlager M. Chromatin signatures of pluripotent cell lines. Nat Cell Biol. 2006;8(5):532.
127. Bernstein BE, Mikkelsen TS, Xie X, Kamal M, Huebert DJ, Cuff J, Fry B, Meissner A, Wernig M, Plath K. A bivalent chromatin structure marks key developmental genes in embryonic stem cells. Cell. 2006;125(2):315–26.
128. Harikumar A, Meshorer E. Chromatin remodeling and bivalent histone modifications in embryonic stem cells. EMBO Rep. 2015;16(12):1609–19.
129. Srinageshwar B, Maiti P, Dunbar GL, Rossignol J. Role of epigenetics in stem cell proliferation and differentiation: implications for treating neurodegenerative diseases. Int J Mol Sci. 2016;17(2):199.
130. Schmidt R, Plath K. The roles of the reprogramming factors Oct4, Sox2 and Klf4 in resetting the somatic cell epigenome during induced pluripotent stem cell generation. Genome Biol. 2012;13(10):251.
131. Vaskova E, Stekleneva A, Medvedev S, Zakian S. "Epigenetic memory" phenomenon in induced pluripotent stem cells. Acta Nat. 2013;5(4):15–21.
132. Nishino K, Toyoda M, Yamazaki-Inoue M, Fukawatase Y, Chikazawa E, Sakaguchi H, Akutsu H, Umezawa A. DNA methylation dynamics in human induced pluripotent stem cells over time. PLoS Genet. 2011;7(5):e1002085.
133. Lowe R, Shirley N, Bleackley M, Dolan S, Shafee T. Transcriptomics technologies. PLoS Comput Biol. 2017;13(5):e1005457.
134. Sengupta S, Prajapati RK, Mukhopadhyay J. Promoter escape with bacterial two-component sigma factor suggests retention of sigma region two in the elongation complex. J Biol Chem. 2015;290:28575–83.
135. ENCODE Project Consortium. An integrated encyclopedia of DNA elements in the human genome. Nature. 2012;489(7414):57.
136. Kukurba KR, Montgomery SB. RNA sequencing and analysis. Cold Spring Harb Protoc. 2015;2015(11):951–69.
137. Rossi S, Calin GA. Bioinformatics, non-coding RNAs and its possible application in personalized medicine. In: MicroRNA cancer regulation. Dordrecht: Springer; 2013. p. 21–37.
138. Stefani G, Slack FJ. Small non-coding RNAs in animal development. Nat Rev Mol Cell Biol. 2008;9(3):219.
139. Natarajan KN, Teichmann SA, Kolodziejczyk AA. Single cell transcriptomics of pluripotent stem cells: reprogramming and differentiation. Curr Opin Genet Dev. 2017;46:66–76.
140. Torres-Padilla M-E, Chambers I. Transcription factor heterogeneity in pluripotent stem cells: a stochastic advantage. Development. 2014;141(11):2173–81.
141. Au KF, Sebastiano V. The transcriptome of human pluripotent stem cells. Curr Opin Genet Dev. 2014;28:71–7.
142. Lockhart DJ, Dong H, Byrne MC, Follettie MT, Gallo MV, Chee MS, Mittmann M, Wang C, Kobayashi M, Norton H. Expression monitoring by hybridization to high-density oligonucleotide arrays. Nat Biotechnol. 1996;14(13):1675.
143. Shalon D, Smith SJ, Brown PO. A DNA microarray system for analyzing complex DNA samples using two-color fluorescent probe hybridization. Genome Res. 1996;6(7):639–45.
144. Djebali S, Davis CA, Merkel A, Dobin A, Lassmann T, Mortazavi A, Tanzer A, Lagarde J, Lin W, Schlesinger F. Landscape of transcription in human cells. Nature. 2012;489(7414):101.
145. Trapnell C, Williams BA, Pertea G, Mortazavi A, Kwan G, Van Baren MJ, Salzberg SL, Wold BJ, Pachter L. Transcript assembly and quantification by RNA-Seq reveals unannotated transcripts and isoform switching during cell differentiation. Nat Biotechnol. 2010;28(5):511.
146. Ahmed T, Begum M. Association between gene expression, clinical factors and survival in patients with breast cancer. J Biomed Anal. 2018;1(1):1–14.

147. Richards M, Tan SP, Tan JH, Chan WK, Bongso A. The transcriptome profile of human embryonic stem cells as defined by SAGE. Stem Cells. 2004;22(1):51–64.
148. van den Berg CW, Okawa S, de Sousa Lopes SMC, van Iperen L, Passier R, Braam SR, Tertoolen LG, del Sol A, Davis RP, Mummery CL. Transcriptome of human foetal heart compared with cardiomyocytes from pluripotent stem cells. Development. 2015;142(18):3231–8.
149. Aslam B, Basit M, Nisar MA, Khurshid M, Rasool MH. Proteomics: technologies and their applications. J Chromatogr Sci. 2017;55(2):182–96.
150. Cox J, Mann M. Is proteomics the new genomics? Cell. 2007;130(3):395–8.
151. Valdespino-Gómez VM, Valdespino-Castillo PM, Valdespino-Castillo VE. Cell signalling pathways interaction in cellular proliferation: potential target for therapeutic interventionism. Cir Cir. 2015;83(2):165–74.
152. Cadeco S, Williamson AJ, Whetton AD. The use of proteomics for systematic analysis of normal and transformed hematopoietic stem cells. Curr Pharm Des. 2012;18(13):1730–50.
153. Iakoucheva LM, Brown CJ, Lawson JD, Obradović Z, Dunker AK. Intrinsic disorder in cell-signaling and cancer-associated proteins. J Mol Biol. 2002;323(3):573–84.
154. Jain KK. Role of proteomics in the development of personalized medicine. Adv Protein Chem Struct Biol. 2016;102:41–52.
155. Midic U, Oldfield CJ, Dunker AK, Obradovic Z, Uversky VN. Protein disorder in the human diseasome: unfoldomics of human genetic diseases. BMC Genomics. 2009;10(1):S12.
156. Baharvand H, Fathi A, van Hoof D, Salekdeh GH. Concise review: trends in stem cell proteomics. Stem Cells. 2007;25(8):1888–903.
157. Unwin RD, Gaskell SJ, Evans CA, Whetton AD. The potential for proteomic definition of stem cell populations. Exp Hematol. 2003;31(12):1147–59.
158. Wang H, Zhang Q, Fang X. Transcriptomics and proteomics in stem cell research. Front Med. 2014;8(4):433–44.
159. Yekta RF, Koushki M, DashAtan NA. Advances in proteomics analytical techniques. J Paramed Sci. 2015;6(3). https://doi.org/10.22037/jps.v6i3.9773.
160. Encarnación-Guevara S. The dawn and the first twenty-five years of proteomics in Mexico: a personal chronicle. Bol Med Hosp Infant Mex. 2017;74(3):208–11.
161. Abazova N, Krijgsveld J. Advances in stem cell proteomics. Curr Opin Genet Dev. 2017;46:149–55.
162. Kovarova H, Gadher SJ, Wollscheid B. Focus on stem cell proteomics. Proteomics. 2011;11(20):3943–5.
163. Mesri M. Advances in proteomic technologies and its contribution to the field of cancer. Adv Med. 2014;2014:238045.
164. Mateos J, Pernas P, Labora J, Blanco F, Arufe M. Proteomic applications in the study of human mesenchymal stem cells. Proteomes. 2014;2(1):53–71.
165. Muñoz J, Heck A Jr. Perspectives in stem cell proteomics. Genome Med. 2009;1(4):45.
166. Heck AJ, Mummery C, Whetton A, Oh S, Lee B, Pera M, Lemischka I, Krijgsveld J. Proteome biology of stem cells. Stem Cell Res. 2007;1(1):7–8.
167. Cho WCS. Proteomics technologies and challenges. Genomics Proteomics Bioinformatics. 2007;5(2):77–85.
168. Miteva YV, Budayeva HG, Cristea IM. Proteomics-based methods for discovery, quantification, and validation of protein-protein interactions. Anal Chem. 2013;85(2):749–68.
169. Vidal BC, Bonventre JV, I-Hong Hsu S. Towards the application of proteomics in renal disease diagnosis. Clin Sci (Lond). 2005;109(5):421–30.
170. Oonk S, Schuurmans T, Pabst M, de Smet LC, de Puit M. Proteomics as a new tool to study fingermark ageing in forensics. Sci Rep. 2018;8(1):16425.
171. Ebhardt HA, Root A, Sander C, Aebersold R. Applications of targeted proteomics in systems biology and translational medicine. Proteomics. 2015;15(18):3193–208.
172. Rual J-F, Venkatesan K, Hao T, Hirozane-Kishikawa T, Dricot A, Li N, Berriz GF, Gibbons FD, Dreze M, Ayivi-Guedehoussou N. Towards a proteome-scale map of the human protein–protein interaction network. Nature. 2005;437(7062):1173.

173. Stevens RC, Yokoyama S, Wilson IA. Global efforts in structural genomics. Science. 2001;294(5540):89–92.
174. Wang J, Trowbridge JJ, Rao S, Orkin SH. Proteomic studies of stem cells. In: StemBook. Cambridge: Harvard Stem Cell Institute; 2008.
175. Coskun O. Separation techniques: Chromatography. North Clin Istanb. 2016;3(2):156–60.
176. Cuatrecasas P, Wilchek M, Anfinsen CB. Selective enzyme purification by affinity chromatography. Proc Natl Acad Sci. 1968;61(2):636–43.
177. Porath J. From gel filtration to adsorptive size exclusion. J Protein Chem. 1997;16(5):463–8.
178. Burgess K, Creek D, Dewsbury P, Cook K, Barrett MP. Semi-targeted analysis of metabolites using capillary-flow ion chromatography coupled to high-resolution mass spectrometry. Rapid Commun Mass Spectrom. 2011;25(22):3447–52.
179. Fiehn O, Kopka J, Dörmann P, Altmann T, Trethewey RN, Willmitzer L. Metabolite profiling for plant functional genomics. Nat Biotechnol. 2000;18(11):1157.
180. McNamara LE, Turner L-A, Burgess KV. Systems biology approaches applied to regenerative medicine. Curr Pathobiol Rep. 2015;3(1):37–45.
181. Jungbauer A, Hahn R. Ion-exchange chromatography. Methods Enzymol. 2009;463:349–71.
182. Ly L, Wasinger VC. Protein and peptide fractionation, enrichment and depletion: tools for the complex proteome. Proteomics. 2011;11(4):513–34.
183. Yamamoto S, Nakanishi K, Matsuno R. Ion-exchange chromatography of proteins. Boca Raton: CRC Press; 1988.
184. Batas B, Chaudhuri JB. Protein refolding at high concentration using size-exclusion chromatography. Biotechnol Bioeng. 1996;50(1):16–23.
185. Hong P, Koza S, Bouvier ES. A review size-exclusion chromatography for the analysis of protein biotherapeutics and their aggregates. J Liq Chromatogr Relat Technol. 2012;35(20):2923–50.
186. Hage DS, Anguizola JA, Bi C, Li R, Matsuda R, Papastavros E, Pfaunmiller E, Vargas J, Zheng X. Pharmaceutical and biomedical applications of affinity chromatography: recent trends and developments. J Pharm Biomed Anal. 2012;69:93–105.
187. Clark MF, Lister RM, Bar-Joseph M. ELISA techniques. Methods Enzymol. 1986;118:742–66.
188. Lequin RM. Enzyme immunoassay (EIA)/enzyme-linked immunosorbent assay (ELISA). Clin Chem. 2005;51(12):2415–8.
189. Kurien BT, Scofield RH. Western blotting. Methods. 2006;38(4):283–93.
190. Hunt DF, Yates JR, Shabanowitz J, Winston S, Hauer CR. Protein sequencing by tandem mass spectrometry. Proc Natl Acad Sci. 1986;83(17):6233–7.
191. Mann M. The rise of mass spectrometry and the fall of Edman degradation. Clin Chem. 2016;62(1):293–4.
192. Rosenberg JM, Utz PJ. Protein microarrays: a new tool for the study of autoantibodies in immunodeficiency. Front Immunol. 2015;6:138.
193. Sutandy FX, Qian J, Chen CS, Zhu H. Overview of protein microarrays. Curr Protoc Protein Sci. 2013;Chapter 27:Unit 27.21.
194. Gorg A, Obermaier C, Boguth G, Harder A, Scheibe B, Wildgruber R, Weiss W. The current state of two-dimensional electrophoresis with immobilized pH gradients. Electrophoresis. 2000;21(6):1037–53.
195. Gorg A, Weiss W, Dunn MJ. Current two-dimensional electrophoresis technology for proteomics. Proteomics. 2004;4(12):3665–85.
196. Marouga R, David S, Hawkins E. The development of the DIGE system: 2D fluorescence difference gel analysis technology. Anal Bioanal Chem. 2005;382(3):669–78.
197. Mohanty AK, Yadav ML, Choudhary S. Gel electrophoresis of proteins and nucleic acids. In: Srivastava N, Pande M, editors. Protocols in semen biology (comparing assays). Singapore: Springer; 2017. p. 233–46.
198. Garman EF. Developments in x-ray crystallographic structure determination of biological macromolecules. Science. 2014;343(6175):1102–8.
199. Smyth M, Martin J. X ray crystallography. Mol Pathol. 2000;53(1):8.

200. Shiio Y, Aebersold R. Quantitative proteome analysis using isotope-coded affinity tags and mass spectrometry. Nat Protoc. 2006;1(1):139.
201. Blueggel M, Chamrad D, Meyer HE. Bioinformatics in proteomics. Curr Pharm Biotechnol. 2004;5(1):79–88.
202. Schmidt A, Forne I, Imhof A. Bioinformatic analysis of proteomics data. BMC Syst Biol. 2014;8(Suppl 2):S3.
203. Di Meo A, Pasic MD, Yousef GM. Proteomics and peptidomics: moving toward precision medicine in urological malignancies. Oncotarget. 2016;7(32):52460–74.
204. Tang HY, Beer LA, Tanyi JL, Zhang R, Liu Q, Speicher DW. Protein isoform-specific validation defines multiple chloride intracellular channel and tropomyosin isoforms as serological biomarkers of ovarian cancer. J Proteome. 2013;89:165–78.
205. Hung MC, Link W. Protein localization in disease and therapy. J Cell Sci. 2011;124(Pt 20):3381–92.
206. Shekari F, Nezari H, Larijani MR, Han CL, Baharvand H, Chen YJ, Salekdeh GH. Proteome analysis of human embryonic stem cells organelles. J Proteome. 2017;162:108–18.
207. Shekari F, Baharvand H, Salekdeh GH. Organellar proteomics of embryonic stem cells. Adv Protein Chem Struct Biol. 2014;95:215–30.
208. Munoz J, Low TY, Kok YJ, Chin A, Frese CK, Ding V, Choo A, Heck AJ. The quantitative proteomes of human-induced pluripotent stem cells and embryonic stem cells. Mol Syst Biol. 2011;7:550.
209. Park IH, Zhao R, West JA, Yabuuchi A, Huo H, Ince TA, Lerou PH, Lensch MW, Daley GQ. Reprogramming of human somatic cells to pluripotency with defined factors. Nature. 2008;451(7175):141–6.
210. Kupcova Skalnikova H. Proteomic techniques for characterisation of mesenchymal stem cell secretome. Biochimie. 2013;95(12):2196–211.
211. Potian JA, Aviv H, Ponzio NM, Harrison JS, Rameshwar P. Veto-like activity of mesenchymal stem cells: functional discrimination between cellular responses to alloantigens and recall antigens. J Immunol. 2003;171(7):3426–34.
212. Arrell DK, Niederlander NJ, Perez-Terzic C, Chung S, Behfar A, Terzic A. Pharmacoproteomics: advancing the efficacy and safety of regenerative therapeutics. Clin Pharmacol Ther. 2007;82(3):316–9.
213. Arjmand B, Goodarzi P, Mohamadi-Jahani F, Falahzadeh K, Larijani B. Personalized Regenerative Medicine. Acta Med Iran. 2017;55(3):144–9.
214. Duarte TT, Spencer CT. Personalized proteomics: the future of precision medicine. Proteomes. 2016;4(4):29.
215. Mathew A, Padmanaban V. Metabolomics: the apogee of the omics trilogy. Int J Pharm Pharm Sci. 2013;5(2):45–8.
216. Trivedi DK, Hollywood KA, Goodacre R. Metabolomics for the masses: the future of metabolomics in a personalized world. New Horiz Transl Med. 2017;3(6):294–305.
217. Cacciatore S, Loda M. Innovation in metabolomics to improve personalized healthcare. Ann N Y Acad Sci. 2015;1346(1):57–62.
218. Fillet M, Frédérich M. The emergence of metabolomics as a key discipline in the drug discovery process. Drug Discov Today Technol. 2015;13:19–24.
219. Schnackenberg LK, Kaput J, Beger RD. Metabolomics: a tool for personalizing medicine. Per Med. 2008;5(5):495–504.
220. Zhang R, Hu P, Zang Q, Yue X, Zhou Z, Xu X, Xu J, Li S, Chen Y, Qiang B. LC-MS-based metabolomics reveals metabolic signatures related to glioma stem-like cell self-renewal and differentiation. RSC Adv. 2017a;7(39):24221–32.
221. Tavassoly I, Goldfarb J, Iyengar R. Systems biology primer: the basic methods and approaches. Essays Biochem. 2018;62(4):487–500.
222. Nicholson J, Lindon JC. Systems biology: Metabonomics. Nature. 2008;455(7216):1054–6.
223. Kitano H. Systems biology: a brief overview. Science. 2002;295(5560):1662–4.

224. Novotny MV, Soini HA, Mechref Y. Biochemical individuality reflected in chromatographic, electrophoretic and mass-spectrometric profiles. J Chromatogr B. 2008;866(1-2):26–47.
225. van der Greef J, Smilde AK. Symbiosis of chemometrics and metabolomics: past, present, and future. J Chemom Soc. 2005;19(5-7):376–86.
226. Pauling L, Robinson AB, Teranishi R, Cary P. Quantitative analysis of urine vapor and breath by gas-liquid partition chromatography. Proc Natl Acad Sci. 1971;68(10):2374–6.
227. Oliver SG, Winson MK, Kell DB, Baganz F. Systematic functional analysis of the yeast genome. Trends Biotechnol. 1998;16(9):373–8.
228. Nicholson JK, Lindon JC, Holmes E. 'Metabonomics': understanding the metabolic responses of living systems to pathophysiological stimuli via multivariate statistical analysis of biological NMR spectroscopic data. Xenobiotica. 1999;29(11):1181–9.
229. Holmes E, Antti H. Chemometric contributions to the evolution of metabonomics: mathematical solutions to characterising and interpreting complex biological NMR spectra. Analyst. 2002;127(12):1549–57.
230. Lenz EM, Wilson ID. Analytical strategies in metabonomics. J Proteome Res. 2007;6(2):443–58.
231. Smith CA, O'Maille G, Want EJ, Qin C, Trauger SA, Brandon TR, Custodio DE, Abagyan R, Siuzdak G. METLIN: a metabolite mass spectral database. Ther Drug Monit. 2005;27(6):747–51.
232. Wishart DS, Mandal R, Stanislaus A, Ramirez-Gaona M. Cancer metabolomics and the human metabolome database. Meta. 2016;6(1):10.
233. Wishart DS, Tzur D, Knox C, Eisner R, Guo AC, Young N, Cheng D, Jewell K, Arndt D, Sawhney S. HMDB: the human metabolome database. Nucleic Acids Res. 2007;35(suppl_1):D521–6.
234. Bingol K. Recent advances in targeted and untargeted metabolomics by NMR and MS/NMR methods. High Throughput. 2018;7(2):9.
235. Roberts LD, Souza AL, Gerszten RE, Clish CB. Targeted metabolomics. Curr Protoc Mol Biol. 2012;98(1):30.2.1–30.2.24.
236. Koen N, Du Preez I, Loots du T. Metabolomics and personalized medicine. Adv Protein Chem Struct Biol. 2016;102:53–78.
237. Panopoulos AD, Yanes O, Ruiz S, Kida YS, Diep D, Tautenhahn R, Herrerías A, Batchelder EM, Plongthongkum N, Lutz M. The metabolome of induced pluripotent stem cells reveals metabolic changes occurring in somatic cell reprogramming. Cell Res. 2012;22(1):168.
238. Puchades-Carrasco L, Pineda-Lucena A. Metabolomics applications in precision medicine: an oncological perspective. Curr Top Med Chem. 2017;17(24):2740–51.
239. West PR, Weir AM, Smith AM, Donley EL, Cezar GG. Predicting human developmental toxicity of pharmaceuticals using human embryonic stem cells and metabolomics. Toxicol Appl Pharmacol. 2010;247(1):18–27.
240. Bekri S. The role of metabolomics in precision medicine. Expert Rev Precis Med Drug Dev. 2016;1(6):517–32.
241. McCartney A, Vignoli A, Biganzoli L, Love R, Tenori L, Luchinat C, Di Leo A. Metabolomics in breast cancer: a decade in review. Cancer Treat Rev. 2018;67:88–96.
242. Smoleńska Ż, Zdrojewski Z. Metabolomics and its potential in diagnosis, prognosis and treatment of rheumatic diseases. Reumatologia. 2015;53(3):152.
243. Ussher JR, Elmariah S, Gerszten RE, Dyck JR. The emerging role of metabolomics in the diagnosis and prognosis of cardiovascular disease. J Am Coll Cardiol. 2016;68(25):2850–70.
244. Beger RD, Dunn W, Schmidt MA, Gross SS, Kirwan JA, Cascante M, Brennan L, Wishart DS, Oresic M, Hankemeier T. Metabolomics enables precision medicine:"a white paper, community perspective". Metabolomics. 2016;12(9):149.
245. Bull MJ, Plummer NT. Part 1: the human gut microbiome in health and disease. Integr Med. 2014;13(6):17.
246. Clish CB. Metabolomics: an emerging but powerful tool for precision medicine. Mol Case Stud. 2015;1(1):a000588.

247. Anderson NM, Mucka P, Kern JG, Feng H. The emerging role and targetability of the TCA cycle in cancer metabolism. Protein Cell. 2018;9(2):216–37.
248. Hallan S, Afkarian M, Zelnick LR, Kestenbaum B, Sharma S, Saito R, Darshi M, Barding G, Raftery D, Ju W, Kretzler M, Sharma K, de Boer IH. Metabolomics and gene expression analysis reveal down-regulation of the citric acid (TCA) cycle in non-diabetic CKD patients. EBioMedicine. 2017;26:68–77.
249. Kuehnbaum NL, Britz-McKibbin P. New advances in separation science for metabolomics: resolving chemical diversity in a post-genomic era. Chem Rev. 2013;113(4):2437–68.
250. Haug K, Salek RM, Conesa P, Hastings J, de Matos P, Rijnbeek M, Mahendraker T, Williams M, Neumann S, Rocca-Serra P. MetaboLights—an open-access general-purpose repository for metabolomics studies and associated meta-data. Nucleic Acids Res. 2012;41(D1):D781–6.
251. Martin J-C, Maillot M, Mazerolles G, Verdu A, Lyan B, Migne C, Defoort C, Canlet C, Junot C, Guillou C. Can we trust untargeted metabolomics? Results of the metabo-ring initiative, a large-scale, multi-instrument inter-laboratory study. Metabolomics. 2015;11(4):807–21.
252. Sansone S-A, Fan T, Goodacre R, Griffin JL, Hardy NW, Kaddurah-Daouk R, Kristal BS, Lindon J, Mendes P, Morrison N. The metabolomics standards initiative. Nat Biotechnol. 2007;25(8):846.
253. Sumner LW, Amberg A, Barrett D, Beale MH, Beger R, Daykin CA, Fan TW-M, Fiehn O, Goodacre R, Griffin JL. Proposed minimum reporting standards for chemical analysis. Metabolomics. 2007;3(3):211–21.
254. Johnson CH, Gonzalez FJ. Challenges and opportunities of metabolomics. J Cell Physiol. 2012;227(8):2975–81.
255. Yin P, Xu G. Current state-of-the-art of nontargeted metabolomics based on liquid chromatography–mass spectrometry with special emphasis in clinical applications. J Chromatogr A. 2014;1374:1–13.
256. Dénes J l, Szabó E, Robinette SL, Szatmári I, Szőnyi L, Kreuder JG, Rauterberg EW, Takáts Z. Metabonomics of newborn screening dried blood spot samples: a novel approach in the screening and diagnostics of inborn errors of metabolism. Anal Chem. 2012;84(22):10113–20.
257. Manwaring V, Boutin M, Auray-Blais C. A metabolomic study to identify new globotriaosylceramide-related biomarkers in the plasma of Fabry disease patients. Anal Chem. 2013;85(19):9039–48.
258. Dunn WB, Broadhurst D, Begley P, Zelena E, Francis-McIntyre S, Anderson N, Brown M, Knowles JD, Halsall A, Haselden JN. Procedures for large-scale metabolic profiling of serum and plasma using gas chromatography and liquid chromatography coupled to mass spectrometry. Nat Protoc. 2011;6(7):1060.
259. Kamleh A, Barrett M, Wildridge D, Burchmore R, Scheltema R, Watson D. Metabolomic profiling using Orbitrap Fourier transform mass spectrometry with hydrophilic interaction chromatography: a method with wide applicability to analysis of biomolecules. Rapid Commun Mass Spectrom. 2008;22(12):1912–8.
260. Gika HG, Wilson ID, Theodoridis GA. LC–MS-based holistic metabolic profiling. Problems, limitations, advantages, and future perspectives. J Chromatogr B. 2014;966:1–6.
261. Tang DQ, Zou L, Yin XX, Ong CN. HILIC-MS for metabolomics: an attractive and complementary approach to RPLC-MS. Mass Spectrom Rev. 2016;35(5):574–600.
262. Ramautar R, Somsen GW, de Jong GJ. CE-MS for metabolomics: developments and applications in the period 2012–2014. Electrophoresis. 2015;36(1):212–24.
263. Mischak H, Coon JJ, Novak J, Weissinger EM, Schanstra JP, Dominiczak AF. Capillary electrophoresis–mass spectrometry as a powerful tool in biomarker discovery and clinical diagnosis: an update of recent developments. Mass Spectrom Rev. 2009;28(5):703–24.
264. Hauschild A-C, Frisch T, Baumbach JI, Baumbach J. Carotta: revealing hidden confounder markers in metabolic breath profiles. Meta. 2015;5(2):344–63.
265. Paglia G, Angel P, Williams JP, Richardson K, Olivos HJ, Thompson JW, Menikarachchi L, Lai S, Walsh C, Moseley A. Ion mobility-derived collision cross section as an additional measure for lipid fingerprinting and identification. Anal Chem. 2014;87(2):1137–44.

266. Kliman M, May JC, McLean JA. Lipid analysis and lipidomics by structurally selective ion mobility-mass spectrometry. Biochim Biophy Acta. 2011;1811(11):935–45.
267. Tebani A, Schmitz-Afonso I, Rutledge DN, Gonzalez BJ, Bekri S, Afonso C. Optimization of a liquid chromatography ion mobility-mass spectrometry method for untargeted metabolomics using experimental design and multivariate data analysis. Anal Chim Acta. 2016;913:55–62.
268. Everett JR. From metabonomics to pharmacometabonomics: the role of metabolic profiling in personalized medicine. Front Pharmacol. 2016;7:297.
269. Urban PL. Quantitative mass spectrometry: an overview. Phil Trans R Soc A. 2016;374:20150382.
270. McNamara LE, Sjöström T, Meek RD, Oreffo RO, Su B, Dalby MJ, Burgess KE. Metabolomics: a valuable tool for stem cell monitoring in regenerative medicine. J R Soc Interface. 2012;9(73):1713–24.
271. Dang TTT, Onoyovwi A, Farrow SC, Facchini PJ. Biochemical genomics for gene discovery in benzylisoquinoline alkaloid biosynthesis in opium poppy and related species. Methods Enzymol. 2012;515:231–66.
272. Newton RP, Brenton AG, Smith CJ, Dudley E. Plant proteome analysis by mass spectrometry: principles, problems, pitfalls and recent developments. Phytochemistry. 2004;65(11):1449–85.
273. Bonhomme C, Gervais C, Laurencin D. Recent NMR developments applied to organic–inorganic materials. Prog Nucl Magn Reson Spectrosc. 2014;77:1–48.
274. Wolfender J-L, Rudaz S, Hae Choi Y, Kyong Kim H. Plant metabolomics: from holistic data to relevant biomarkers. Curr Med Chem. 2013;20(8):1056–90.
275. Bingol K, Brüschweiler R. Multidimensional approaches to NMR-based metabolomics. Anal Chem. 2013;86(1):47–57.
276. Powers R. The current state of drug discovery and a potential role for NMR metabolomics: miniperspective. J Med Chem. 2014;57(14):5860–70.
277. Rankin NJ, Preiss D, Welsh P, Burgess KE, Nelson SM, Lawlor DA, Sattar N. The emergence of proton nuclear magnetic resonance metabolomics in the cardiovascular arena as viewed from a clinical perspective. Atherosclerosis. 2014;237(1):287–300.
278. Dunn WB, Bailey NJ, Johnson HE. Measuring the metabolome: current analytical technologies. Analyst. 2005;130(5):606–25.
279. Halket JM, Waterman D, Przyborowska AM, Patel RK, Fraser PD, Bramley PM. Chemical derivatization and mass spectral libraries in metabolic profiling by GC/MS and LC/MS/MS. J Exp Bot. 2004;56(410):219–43.
280. Schattka B, Alexander M, Ying SL, Man A, Shaw RA. Metabolic fingerprinting of biofluids by infrared spectroscopy: modeling and optimization of flow rates for laminar fluid diffusion interface sample preconditioning. Anal Chem. 2010;83(2):555–62.
281. Lasch P, Chiriboga L, Yee H, Diem M. Infrared spectroscopy of human cells and tissue: detection of disease. Technol Cancer Res Treat. 2002;1(1):1–7.
282. Huang H-M, Shih Y-Y. Pushing CT and MR imaging to the molecular level for studying the "omics": current challenges and advancements. Biomed Res Int. 2014;2014:365812.
283. Beckonert O, Coen M, Keun HC, Wang Y, Ebbels TM, Holmes E, Lindon JC, Nicholson JK. High-resolution magic-angle-spinning NMR spectroscopy for metabolic profiling of intact tissues. Nat Protoc. 2010;5(6):1019.
284. Cacciatore S, Hu X, Viertler C, Kap M, Bernhardt GA, Mischinger HJ, Riegman P, Zatloukal K, Luchinat C, Turano P. Effects of intra-and post-operative ischemia on the metabolic profile of clinical liver tissue specimens monitored by NMR. J Proteome Res. 2013;12(12):5723–9.
285. Shyh-Chang N, Ng H-H. The metabolic programming of stem cells. Genes Dev. 2017;31(4):336–46.
286. Bhute VJ, Bao X, Palecek SP. Advances in applications of metabolomics in pluripotent stem cell research. Curr Opin Chem Eng. 2017;15:36–43.
287. Shyh-Chang N, Daley GQ, Cantley LC. Stem cell metabolism in tissue development and aging. Development. 2013a;140(12):2535–47.

288. Riester M, Xu Q, Moreira A, Zheng J, Michor F, Downey R. The Warburg effect: persistence of stem-cell metabolism in cancers as a failure of differentiation. Ann Oncol. 2017;29(1):264–70.
289. Vander Heiden MG, Cantley LC, Thompson CB. Understanding the Warburg effect: the metabolic requirements of cell proliferation. Science. 2009;324(5930):1029–33.
290. Papa S, Martino PL, Capitanio G, Gaballo A, De Rasmo D, Signorile A, Petruzzella V. The oxidative phosphorylation system in mammalian mitochondria. In: Advances in Mitochondrial Medicine. Dordrecht: Springer; 2012. p. 3–37.
291. Ito K, Suda T. Metabolic requirements for the maintenance of self-renewing stem cells. Nat Rev Mol Cell Biol. 2014;15(4):243.
292. Pantaleon M, Kaye PL. Glucose transporters in preimplantation development. Rev Reprod. 1998;3(2):77–81.
293. Varum S, Rodrigues AS, Moura MB, Momcilovic O, Easley CA IV, Ramalho-Santos J, Van Houten B, Schatten G. Energy metabolism in human pluripotent stem cells and their differentiated counterparts. PLoS One. 2011;6(6):e20914.
294. Folmes CD, Nelson TJ, Martinez-Fernandez A, Arrell DK, Lindor JZ, Dzeja PP, Ikeda Y, Perez-Terzic C, Terzic A. Somatic oxidative bioenergetics transitions into pluripotency-dependent glycolysis to facilitate nuclear reprogramming. Cell Metab. 2011;14(2):264–71.
295. Shyh-Chang N, Locasale JW, Lyssiotis CA, Zheng Y, Teo RY, Ratanasirintrawoot S, Zhang J, Onder T, Unternaehrer JJ, Zhu H. Influence of threonine metabolism on S-adenosylmethionine and histone methylation. Science. 2013b;339(6116):222–6.
296. Yoshida Y, Takahashi K, Okita K, Ichisaka T, Yamanaka S. Hypoxia enhances the generation of induced pluripotent stem cells. Cell Stem Cell. 2009;5(3):237–41.
297. Balzer E, Moss EG. Localization of the developmental timing regulator Lin28 to mRNP complexes, P-bodies and stress granules. RNA Biol. 2007;4(1):16–25.
298. Peng S, Chen LL, Lei XX, Yang L, Lin H, Carmichael GG, Huang Y. Genome-wide studies reveal that Lin28 enhances the translation of genes important for growth and survival of human embryonic stem cells. Stem Cells. 2011;29(3):496–504.
299. Viswanathan SR, Daley GQ, Gregory RI. Selective blockade of microRNA processing by Lin28. Science. 2008;320(5872):97–100.
300. Sun X, Liu J, Xu C, Tang S-C, Ren H. The insights of Let-7 miRNAs in oncogenesis and stem cell potency. J Cell Mol Med. 2016;20(9):1779–88.
301. Zhu H, Shyh-Chang N, Segrè AV, Shinoda G, Shah SP, Einhorn WS, Takeuchi A, Engreitz JM, Hagan JP, Kharas MG. The Lin28/let-7 axis regulates glucose metabolism. Cell. 2011;147(1):81–94.
302. Shyh-Chang N, Zheng Y, Locasale JW, Cantley LC. Human pluripotent stem cells decouple respiration from energy production. EMBO J. 2011;30(24):4851–2.
303. Manganelli G, Fico A, Masullo U, Pizzolongo F, Cimmino A, Filosa S. Modulation of the pentose phosphate pathway induces endodermal differentiation in embryonic stem cells. PLoS One. 2012;7(1):e29321.
304. Gaspar-Maia A, Alajem A, Meshorer E, Ramalho-Santos M. Open chromatin in pluripotency and reprogramming. Nat Rev Mol Cell Biol. 2011;12(1):36.
305. Huangfu D, Maehr R, Guo W, Eijkelenboom A, Snitow M, Chen AE, Melton DA. Induction of pluripotent stem cells by defined factors is greatly improved by small-molecule compounds. Nat Biotechnol. 2008;26(7):795.
306. Moussaieff A, Rouleau M, Kitsberg D, Cohen M, Levy G, Barasch D, Nemirovski A, Shen-Orr S, Laevsky I, Amit M. Glycolysis-mediated changes in acetyl-CoA and histone acetylation control the early differentiation of embryonic stem cells. Cell Metab. 2015;21(3):392–402.
307. Mohyeldin A, Garzón-Muvdi T, Quiñones-Hinojosa A. Oxygen in stem cell biology: a critical component of the stem cell niche. Cell Stem Cell. 2010;7(2):150–61.
308. Suda T, Takubo K, Semenza GL. Metabolic regulation of hematopoietic stem cells in the hypoxic niche. Cell Stem Cell. 2011;9(4):298–310.

309. Spencer JA, Ferraro F, Roussakis E, Klein A, Wu J, Runnels JM, Zaher W, Mortensen LJ, Alt C, Turcotte R. Direct measurement of local oxygen concentration in the bone marrow of live animals. Nature. 2014;508(7495):269.
310. Eliasson P, Jönsson JI. The hematopoietic stem cell niche: low in oxygen but a nice place to be. J Cell Physiol. 2010;222(1):17–22.
311. Gupta N, Nizet V. Stabilization of hypoxia-inducible factor 1-alpha augments the therapeutic capacity of bone marrow-derived mesenchymal stem cells in experimental pneumonia. Front Med. 2018;5:131.
312. Rossi DJ, Jamieson CH, Weissman IL. Stems cells and the pathways to aging and cancer. Cell. 2008;132(4):681–96.
313. Pattappa G, Thorpe SD, Jegard NC, Heywood HK, de Bruijn JD, Lee DA. Continuous and uninterrupted oxygen tension influences the colony formation and oxidative metabolism of human mesenchymal stem cells. Tissue Eng Part C Methods. 2012;19(1):68–79.
314. Chung S, Dzeja PP, Faustino RS, Perez-Terzic C, Behfar A, Terzic A. Mitochondrial oxidative metabolism is required for the cardiac differentiation of stem cells. Nat Rev Cardiol. 2007;4(S1):S60.
315. Renault VM, Rafalski VA, Morgan AA, Salih DA, Brett JO, Webb AE, Villeda SA, Thekkat PU, Guillerey C, Denko NC. FoxO3 regulates neural stem cell homeostasis. Cell Stem Cell. 2009;5(5):527–39.
316. Tothova Z, Kollipara R, Huntly BJ, Lee BH, Castrillon DH, Cullen DE, McDowell EP, Lazo-Kallanian S, Williams IR, Sears C. FoxOs are critical mediators of hematopoietic stem cell resistance to physiologic oxidative stress. Cell. 2007;128(2):325–39.
317. Simsek T, Kocabas F, Zheng J, DeBerardinis RJ, Mahmoud AI, Olson EN, Schneider JW, Zhang CC, Sadek HA. The distinct metabolic profile of hematopoietic stem cells reflects their location in a hypoxic niche. Cell Stem Cell. 2010;7(3):380–90.
318. Ito K, Carracedo A, Weiss D, Arai F, Ala U, Avigan DE, Schafer ZT, Evans RM, Suda T, Lee C-H. A PML–PPAR-δ pathway for fatty acid oxidation regulates hematopoietic stem cell maintenance. Nat Med. 2012;18(9):1350.
319. Ito K, Turcotte R, Cui J, Zimmerman SE, Pinho S, Mizoguchi T, Arai F, Runnels JM, Alt C, Teruya-Feldstein J. Self-renewal of a purified Tie2+ hematopoietic stem cell population relies on mitochondrial clearance. Science. 2016;354(6316):1156–60.
320. Fukawa T, Yan-Jiang BC, Min-Wen JC, Jun-Hao ET, Huang D, Qian C-N, Ong P, Li Z, Chen S, Mak SY. Excessive fatty acid oxidation induces muscle atrophy in cancer cachexia. Nat Med. 2016;22(6):666.
321. Klimmeck D, Hansson J, Raffel S, Vakhrushev SY, Trumpp A, Krijgsveld J. Proteomic cornerstones of hematopoietic stem cell differentiation: distinct signatures of multipotent progenitors and myeloid committed cells. Mol Cell Proteomics. 2012;11(8):286–302.
322. Takubo K, Nagamatsu G, Kobayashi CI, Nakamura-Ishizu A, Kobayashi H, Ikeda E, Goda N, Rahimi Y, Johnson RS, Soga T. Regulation of glycolysis by Pdk functions as a metabolic checkpoint for cell cycle quiescence in hematopoietic stem cells. Cell Stem Cell. 2013;12(1):49–61.
323. Chen CT, Shih YRV, Kuo TK, Lee OK, Wei YH. Coordinated changes of mitochondrial biogenesis and antioxidant enzymes during osteogenic differentiation of human mesenchymal stem cells. Stem Cells. 2008;26(4):960–8.

Genomics, Proteomics, and Metabolomics for Stem Cells Monitoring in Regenerative Medicine

Saeed Heidari-Keshel, Azam Rahimi, Mostafa Rezaei-Tavirani, Farshid Sefat, and Arash Khojasteh

Abstract Stem cells are undifferentiated cells that contain long-term potency for differentiation and self-renewal. There is a fine interest to know the mechanisms for stem cells behavior, and use their capacity in medicine researches, developmental and aging studies. In addition to growth factors and morphogens, multiple metabolic pathways take part in the stem cell fate regulation. In current chapter we aim to discuss about stem cell metabolomics such as the ways to maintain stem cell in proliferation and quiescence mood in the hypoxic and normoxic niche. We then describe the mechanisms by which stem cells retain their multipotency properties. Several mechanisms such as oxidative stress which involve in aging reviewed below to explain the decrease in stem cells numbers and function. Then we will go to discuss about the genomics of stem cells. The regulatory genome is significantly important in maintaining pluripotency, therefore scientists attempt to find other components that involved in stem cells differentiation or maintaining self-renewal. We will then explain the specific features of the genomics of IPCs and ESCs. The present work aimed to use the newest genetic engineering techniques combined with in vitro and in vivo imaging applications to realize the full translational potential of hESCs and iPSCs.

Keywords Genomics · Proteomics · Metabolomics · Regenerative medicine

---

S. Heidari-Keshel · A. Rahimi · A. Khojasteh
Department of Tissue Engineering and Applied Cell Sciences, School of Advanced Technologies in Medicine, Shahid Beheshti University of Medical Sciences, Tehran, Iran

M. Rezaei-Tavirani (✉)
Proteomics Research Center, Shahid Beheshti University of Medical Sciences, Tehran, Iran

F. Sefat
Biomedical and Electrical Engineering Department, School of Engineering, University of Bradford, Bradford, UK

Interdisciplinary Research Centre in Polymer Science & Technology (IRC Polymer), University of Bradford, Bradford, UK

B. Arjmand (ed.), *Genomics, Proteomics, and Metabolomics*, Stem Cell Biology and Regenerative Medicine, https://doi.org/10.1007/978-3-030-27727-7_2

1 Introduction

Regenerative medicine as a new approach has attracted attention of researchers and physicians to promote therapeutic protocols. Stem cells play crucial roles in regenerative medicine, due to maintenance and regeneration of tissue and reducing functional cell degeneration [1].

Stem cells of multicellular organisms are undifferentiated unique cells that possess long-term competence for self-renewal and differentiation. There is a tremendous interest in comprehension of the mechanisms for stem cells behavior, and their competence usage in medicine researches as well as developmental and ageing studies. Recent progression in metabolomics sciences has developed our comprehension of specification and self-renewal of stem cells. Therefore, in addition to growth factors and morphogenesis, several metabolic pathways participate in the arrangement of stem cell fate. Various metabolites regulate epigenetic alterations, such as acetylation and histone methylation. These changes are definitive regulators of stem cell destination, in addition to their regulation of oxidative phosphorylation and glycolysis variability during their life span [2]. Metabolomics can also change cues in the extrinsic conditions to regulate cell fates. Stem cell metabolism demonstrates a balance between intrinsic metabolic requirements and extrinsic limitation. Recent studies discussed the outcome of metabolism on the stem cells fate and maintenance as well as the effects of metabolism on proliferation in stem cells and ageing signaling mechanisms in stem cells. In the base of tissue engineering and regenerative medicine, cell metabolomics has a remarkable advantage due to the proteomics, the other significant functional-level tool for cells and material analysis, in that metabolites can be detected with sample material [3].

Proteomics studies proteome of biological sample and proteome as a dynamic subject which includes all expressed proteins in the desired sample such as organelle, cell, tissue, or organism. In this regard expression, post-translational modification, protein–protein interaction networks, biological processes, and biochemical pathway analysis reveal clear perspective of living system and abnormalities [4–7]. Proteomics history is tied to two-dimensional gel electrophoresis and mass spectrometry. However, gel free proteomics and advanced technology are applied in proteomics to analyze biological sample in more details today. Like proteomics, metabolomics is concerned with metabolome which includes whole metabolite indexes in a cell or tissue under fitting situations [8, 9].

It supplements the other "omics" technologies such as genomics and transcriptomics to present the exploration of proteins of an organism, and to realize the structure and functions of a particular protein. Proteomics-based technologies are employed in a variety of potentials for different experiments settings such as finding several diagnostic markers, vaccine production, understanding the pathogenicity mechanisms, alteration in the pattern of expression in response to different signals, and explanation of functional protein pathways in different diseases [10]. These strategies are perfected and the mixture of protein data give raise to proteomic procedure, and the data come from metabolic analysis simplify the universal conception of variation in profiles of stem cells both during proliferation and differentiation

[11]. Genomics is defined as a study of genes and their functions and related technology in the 2002 report of World Health Organization.

Recent advances of human PSC technology happen together with recent development in genomics technologies. Genetic and epigenetic data have prepared valuable vision into the genetic background of recognized genetic and epigenetic varieties associated with disorders. Unfortunately, most of the alternatives have no clear clinical relevance [11]. The technologies of single-cell genomics are being used to engineer tissues to produce cell diversity atlases in these systems [12]. It seems that "omics" studies have high impact to improve regenerative researches and applications. Here metabolomic, proteomic, and genomic aspects of stem cells as fundamental elements in regenerative medicine are presented.

2 Stem Cell Metabolomics

2.1 Somatic Stem Cell Metabolomics

2.1.1 Maintaining Adult/Somatic Stem Cell in Quiescence or Proliferation Mood

2.1.1.1 Hypoxic Niche

Stem cells have rather different innate metabolic needs, unlike the stem cells are hyper proliferative, the majority of adult stem cells stay quiescent in their niches unless stress or injury occurs. Stem cells are usually deep in the tissue, where the hypoxic conditions required for normal stem cell metabolism are provided. For example, neural stem cells inhabit in a brain's area named sub-ventricular zone, satellite cells reside beneath the basal lamina of skeletal muscles, and hematopoietic stem cells in the bone marrow [13–15]. For the purpose of tissue maintenance, stem cells retain a quiescent mood to maintain self-renewal in long term. Quiescent condition in the adult stem cells is generally interdependent to hypoxic niche and glycolytic metabolism. The metabolism of adult stem cells leads to quiescence and proliferation depends on oxygen availability. In hypoxic niches quiescent stem cells prefer to do glycolysis and oxidation which suppress ROS signaling but proliferative stem cells prefer to be in a normoxic environment and use more oxygen and activate ROS signaling [16–20].

2.1.1.2 ROS Signals

ROS is a stress signal that is necessary for cellular processes in normal conditions. It can arise from various sources, such as cytosolic oxygenases and oxidase, incompetent electron transfer by OxPhos complexes in mitochondria or exposure to radiation, and other kind of pollutants that elevate the rate of antioxidant enzymes.

Activation of mitochondrial OxPhos and ROS levels increase when HSC start proliferating, resulting in HSC differentiation. Recent researches suggested that p38 MAPK can be another ROS target that activates proliferation of adult stem cell [21–23]. In the mitochondria for catabolizing fatty acids, fatty acid oxidation or β-oxidation is the series of redox reactions that are significantly important. FAO blockage or inhibition of the upstream regulators lead to symmetric divisions in differentiating cells into committed progenitors, whereas activation of upstream regulators increased self-renewal and lead to asymmetric divisions. It was indicated that FAO flux is essential to symmetric differentiation in NSC self-renewal [24].

2.1.2 Maintaining Adult/Somatic Stem Cell Pluripotent

2.1.2.1 OxPhos in Pluripotency

OxPhos refers to a set of reactions in mitochondrial electron transporting chain which produce ATP using the energy generate by the oxidation. Which is in turn obtain from nutrients oxidation through redox reactions. It is the most competent source of energy for cells, although as the alternative source, some cell types rely on substrate phosphorylation. During mammalian development in the preimplantation phase pluripotency appears in the epiblast. The naïve phase is related to pluripotent stem cells in the embryos in preimplantation state that becomes primed later [25, 26]. PSCs display significant differences in their generation of germline-competent stem cells, epigenomic states, lineage-specific markers, and gene expression of pluripotency markers, signaling necessities to preserve self-renewal and carbon metabolism. Naïve PSCs consume more OxPhos particularly, but primed PSCs depend on glycolysis entirely. In fact OxPhos in PSCs relies on the medium culture conditions and composition, but no intrinsic changes appear in the pluripotency state [27]. Increasing evidence supports the shift in bio-energetic metabolism as an idea is arranged by pluripotency factors intrinsically, and regulates epigenetic machinery that is associated with the programming of the primed and naïve pluripotency phases [28, 29]. The Krebs cycle intermediates are essential for modifying epigenetic states of both differentiating and naïve pluripotency stem cells. This can be a reason why both differentiating PSCs and naïve PSCs promote mitochondrial OxPhos, while primed PSCs rely on glycolysis. More researches are required to entirely clarify the roles of various modes of pluripotency metabolism. Moreover, the detailed culturing conditions for naïve PSCs are controversial, therefore their metabolic requires are still not completely clarified [30].

2.1.2.2 Glycolysis in Reprogramming and Pluripotency

Glycolysis is the redox reaction occurs in the cytosol to produce carbon pyruvate and ATP molecules via phosphorylation. The intermediates may also be diverted into macromolecule synthesis during quick cell growth. Glycolysis able to generate ATP and anabolic growth intermediates rapidly owing to the higher rate of glycolysis reactions. In transitioning from the primary to the differentiating pluripotent state, PSCs reduce the rates of OxPhos intrinsically and have high levels of glycolysis according to expression of transporter of glucose, which leads to high level of glycolysis and glucose uptake. Furthermore, it has been demonstrated that in PSCs, high levels of glucose are used to synthesize nucleotides. Hence, it is shown that the high level of glycolysis and glucose uptake is essential to meet the requirement of proliferation of PSCs [31–33]. The somatic cells reprogramming into iPSCs suggest another way to the metabolic needs of pluripotency. A shift from OxPhos to glycolysis is seen widely in iPSC reprogramming. It is observed that these metabolisms play a significant role in reprogramming. It has been noted that an early mitochondria fragmentation occurs. The mitochondrial fission depends on Drp1 which is necessary for promoting pluripotency. Biogenesis in mitochondria is also critical in reprogramming regulation [34]. Human naïve PSCs display increased glycolysis in comparison to primed PSCs. In naïve PSCs the induction of glycolysis is related to nuclear C-MYC and N-MYC. It is important to note that the existence of feeder-secreted molecule affects the association of glucose with self-renewal of primed PSCs. Therefore it is known that glycolysis has a deep effect on the pluripotent phase of PSCs. Additional work may be required to find out how glycolysis controls differentiation of lineage-specific cells and exits from pluripotency [35]. In the self-renewal PSCs, maintaining in pluripotency and undifferentiated state Glycine–methionine metabolism also has an important role. Moreover, a significant up-regulation of Thr dehydrogenase is found in PSCs in contrast to differentiated cells. However, the being particular of Thr deprivation to PSCs propose that the metabolites that ensue from degradation of Thr-Gly can be used to preserve the pluripotent state. These research have demonstrated the magnitude of Gly-Met metabolism to specify the PSCs cell fate. Histone acetylation leads to the open euchromatin state that holds the pluripotent state epigenetically and PSCs self-renewal. It is important to note that chemical reduction of histone deacetylases induce somatic cells reprogramming into iPSCs [36].

2.1.3 Adult Stem Cell in Aging

After several proliferation periods, adult stem cells including epidermal stem cells, HSCs, and NSCs gently decrease in their numbers, self-renewal, and multipotent potential for differentiation. This continuous decline associates with several degenerative states of ageing such as hair loss and neurodegenerative diseases. Several mechanisms which reviewed below have been involved to explain this decrease in numbers and function [37].

2.1.3.1 Oxidative Stress

The insulin–PI3K–AKT pathway prevents the FoxO transcription factors. Excessive ROS lead to oxidative response by activating the FoxO. Deficiency in the FoxO transcription factors compromises the oxidative stress response and gradually depletes mouse LT-HSCs during ageing [38–41].

2.1.3.2 Metabolic Programming

Studies explained the role of the oxidative stress response regulated with FoxO by showing that p53 and PI3K signaling, ending in a vicious cycle that result to HSC ageing. This defective cycle can be cut off by antioxidant molecules administration. Therefore regulating ROS and FoxOs also activate autophagy and mitophagy to support HSCs during dietary limitations. Mitochondrial metabolism may be regulated by FoxO3. It was demonstrated that germline stem cells in Drosophila and Caenorhabditis elegans regulate their proliferation relevant to nutritional conditions and of course in response to insulin–PI3K–FoxO pathway which is the ageing regulatory signaling [42].

2.1.3.3 AMPK Signaling

ATP/AMP pathway also increased organismal life span by adjusting mitochondrial FAO and repressing mTOR signaling. Defection of LKB1 which is AMPK regulatory factor results in mitochondrial dysfunction, direct to LT-HSC proliferation and defects in hematopoiesis during ageing period and also regulates NSCs and neurons mitochondria. AMPK signaling in addition regulates the quiescent germline stem cells to effect on nematode ageing. The variation in stem cell models has indicated that mitochondrial OxPhos regulates stem cell ageing, and therefore their life span [43]. The mTOR signaling pathway also plays a significant role in modulating mitophagy and adjusting mitochondrial activity to restrain stem cell ageing [44].

2.2 *ESCs and iPSCs Metabolomics*

The high grade of resemblance between iPSCs and ESCs has been indicated on several levels such as developmental capacity and pluripotency. Gene expression assays, such as profiling and analysis of ESC genes, and epigenetic studies trusted high degree of iPSCs and ESCs similarity [45–47]. Most of the reports noted some differences between these pluripotent stem cells, elucidating that while the two cell types are similar, therefore they are not equal. Recent research reported the rejection of iPSCs derived teratomas in contrast to ESCs derived teratomas, both from mouse. Furthermore, iPSCs may contain mutations of unknown function. Cellular metabolism is an important and of course less explored area depending on the pluripotency and biology of stem cells [48]. Metabolomics advanced our knowledge about one type of stem cells as well as their similarity of other stem cell types, as it significantly characterizes the chemical and biological state of a specific cell type. The metabolome includes a great number of components related by multiplex pathways spanning an extensive array of biochemical and structural classes including amino acids, carbohydrates, lipids, and nucleotides. This chemical variety presents a significant challenge so that analytical strategies cannot encompass such various biochemical metabolites. Scientists used four metabolomics programs for analysis: nanoESIMS or nano electrospray ionization mass spectrometry, GCTOF MS or gas chromatography time-of-flight mass spectrometry, HILIC or hydrophilic interaction, and HILIC/RP-QTOF MS or reverse phase chromatography quadrupole time-of-flight mass spectrometry, to support a greater width of the various stem cell metabolome [49]. The data integrated from platforms and graphs cover the notion that near full metabolome reprogramming happened in iPSCs. In addition, there are some important metabolic alteration between iPSCs and ESC. They differ in metabolites of polyamine biosynthesis as well as amino acids and lipid profiles. The iPSCs demonstrate several diversities from the m15 fibroblast cells that were not discovered in the comparison of mESC/m15. Purine metabolism was discovered to be remarkably changed in iPSCs with a reduction of the adenosine-5-phosphate and its deamination product which led to a growth in xanthine which is a metabolite in the pathway of purine salvage [50]. In addition to the more pronounced reduction in amino acid metabolism, reprogramming in iPSCs metabolism had significant effects on purine and polyamine biosynthetic pathways as nitrogen metabolism pathways. The iPSCs and mESCs were directly compared to each other to result in a better prospect of changes in profiles of metabolite. In spite of overall similarity between iPSCs and mESCs, some differences were clear between these two stem cell lines. Both iPSCs and mESCs represent the same lipid profiles but the differences are significantly various by cell type. Again, the appearance of 5-methylthioadenosine and putrescine implicated an effect on the polyamine biosynthesis which led to an increase in xanthine, a metabolite in the purine salvage pathway [51]. In addition to reduction in amino acid metabolism, reprogramming in iPSCs metabolism had essential effects on nitrogen pathways. The iPSCs and mESCs comparison helps to obtain a complete overview of changes in metabolomics [52–54].

3 Stem Cell Genomics

3.1 Regulatory Genome

Coding sequences as well as genome consist of regulatory architecture named cis-regulatory modules which contain a series of TF-binding sites placed in the presence of the regulated gene. As previously explained, differentiation needs to establish a novel regulatory transcriptional state. The devices for information processing into the genome read the defined transcriptional regulatory states as well as process the information then guide the critical transcription apparatus in order to turn on or off the specific genes. The human stem cells genome is a series of instructions that encode proteins, non-coding RNAs as well as how much and when each of these will be expressed in order to produce the multiplex body plan of an organism. Some regulatory genome examples in stem cells are the regulatory morphogens such as Nanog, Oct4, Klf4, and Sox2 in the embryonic stem cells in which activation of them is necessary for retaining pluripotency in stem cells. This regulatory genome is important in maintaining pluripotency, so that scientists are trying to find other novel components that involved in the leading stem cells to differentiation or maintaining self-renewal [55, 56].

3.2 *Epigenetic Aspects of Stem Cell Differentiation*

Epigenetic mechanisms supply both stable and flexible mode of gene regulation needed for development. Histone modifications allow short-term as well as flexible regulation whereas DNA methylation allows long-term as well as stable silencing. Therefore many epigenomes can regulate the entire genome. In addition, the genetic templates dynamically produced with the mixture of genome and epigenomes that prepper the novel gene expression profiles. Epigenetic data is heritable information which is not encoded in the sequence of DNA but results in gene expression. DNA packaging histones or covalent modifications store epigenetic information. Hence, an epigenome is determined as a modificatory factor that binds to chromatin [57]. DNA methylation directs to DNA modification by cytosine methylation, which occurs in CpG dinucleotides. Recent findings indicate that epigenome is essential for successful cloning which supports DNA methylation that provides a novel cellular function in development. Two major mechanisms can influence gene transcription. Firstly, methylation in site-specific DNA can participate in binding to transcription factors of DNA that promote gene transcription. Secondly, methylation in methyl-CpG-binding proteins is associated with transcriptional silencing. DNA methylation is associated with gene expression long-term silencing, which plays a significant role in the control of pluripotency genes. Therefore, an epigenetic perspective in stem cell differentiation can provide new vision on carcinogenesis. Detailed analysis of methylation in DNA supports pluripotency process in iPSCs as well as in ES cells. Epigenetic data is stored by chemical modifications of histones, including ubiquitination, phosphorylation, acetylation, and methylation. A few reports explain histone acetylation as significant in stem cell pluripotency. This epigenomic regulation focuses on stem cell studies, with DNA methylation to support self-renewal and control differentiation [58, 59].

3.3 *iPCs and ESCs Genomics*

One possible definition for the genomic instability observed in iPSCs is conversion in the honesty of pathways involved in DNA repair. The recent researches showed that iPSCs are probably more than other stem cell types to present genomic abnormalities as well as lower fidelity in genome damage repair. They also found to have more DNA stability in iPSCs than iPSCs. The latest engineering techniques in genetics combined with imaging applications are used to find out the translational potential of iPSCs and hESCs. This is the first usage of genome editing for molecular imaging of iPSCs and hESCs. Both iPSCs and hESCs edited and attain a high yield of integration. It is interesting to note that, due to recent advances in generating stem cell in the clinic, recent data have significant implications for the genetic engineering. Transgenesis methods are highly dependent on fundamental limitations. For this purpose, researchers believe that addition of isogenic-targeted to a safe port is the favorite technology for science in clinical applications or preclinical. According to the last reports human iPSC technology has removed the technical and of course ethical issues in producing disease models and has quickly evolved recently. After in vitro differentiation, human iPSCs derived from patient provides somatic cells that are disease-relevant, which carry whole genetic elements involved in the disease, showing the genetic spectrum of population. Recent studies described how human iPSCs can bridge the gaps by producing models related to disease from the patient cells. Noted human iPSCs approach let to simulate patient context in vivo and disease platform. Novel disease hypotheses can be generated according to the particular clinical observations and test them functionally with molecular precision in genomics models or use insights attained from genomics models and support the relevance in the context of patients. In order in the continuing development in tissue engineering and ability to produce organoid structures similar to entire organs, scientists suggested that patient-derived model will become an important research tool in our knowledge of physiology and disease advancement seen in model organisms [60–62].

4 Stem Cell Proteomics

4.1 A Perspective

Proteomics based on 2-D gel electrophoresis, X-ray crystallography, mass spectrometry, and chromatography techniques is applied to study protein expression, function, and structure in the stem cell field [63–65]. Improvement of proteomics technology elevated studies of stem cells. Diversity of stem cells and comparison of stem cells to the other types of cells attracted attention of researchers to provide precious characters of stem cells and to apply in regenerative medicine [66].

In recent times, stem cell activities are known mostly via merging transcriptome and proteome findings. Roche et al. and Park et al. investigated important analysis about mesenchymal stem cells. Van Hoof has represented valuable evaluation about proteomic characteristics of human ESCs [67–70]. There are appreciated information about proteome, genome, and transcriptome of neural, embryonic, and hematopoietic stem cells via combined data from high throughput technologies which deals with [71, 72].

4.2 Proteomic Studies of Embryonic Stem Cells (ESCs)

Suitable characterization of ESCs requires doing wide molecular investigation to ensure safety to apply these key cells in regenerative medicine. Membrane proteins as cell surface markers are used for sorting or isolation of hESC proteins. These proteins also act as extracellular receptors in membrane of stem cells that play roles in cell signaling and regulation of substances transport. Other types of cellular and subcellular proteins that may be about handers or thousands numbers of related proteins in hESCs can be determined via proteomics to analyze biological processes and biochemical pathways that differentiate stem cells into desired cell types [73–76].

Extraordinarily, some proteins such as transcription factor Sox2 are associated with undifferentiated hESC and are known as crucial ones in self-renewal of stem cells [50]. Oct4 the other transcription factor plays critical role in differentiation of mESCs into primitive endoderm and mesoderm [77–80]. It is clear that regulation of protein expression of several proteins controls stemness in ESCs.

Proteomic study revealed that feeder layer plays important role to support undifferentiated generations of hESCs via several proteins which are known as secreted intracellular proteins.

Due to PTMs occurrences, mRNA profile change and proteomic findings provide complimentary information about true dynamics of protein activity and molecular events in stem cells [81, 82]. Studies of PTMs provide new conception of overall molecular mechanism which are involved in system function in hESCs.

4.3 Proteomic Studies of Adult Stem Cells

The other types of stem cells are known as adult stem cells that as well as mesenchymal stem cells (MSCs) are characterized by their developmental ability. However, MSCs were first separated from bone-marrow but they are observed in various tissues such as skin, muscle, tendon, adipose tissue, lungs, dental pulp, placenta, and umbilical cord. It is confirmed that MSC can be differentiated into mesoderm-derived tissues such as osteocytes, adipocytes, and chondrocytes [83–85].

Based on merged datasets from proteomic profiling studies it is reported that largest conserved groups of proteins in adult stem cells probably are involved in energy metabolism. Using sensitive iTRAQ methodology composed of MS analysis showed that hematopoietic stem cells are adapted for anaerobic environments. 2-dimensional liquid chromatography or LC/LC fragmentation followed by tandem MS/MS utilizing revealed that osteogenic differentiation results from the centralizing of gene expression in functional groups rather than simply from the increased expression of new genes. Effects of PTMs (such as tyrosine phosphorylation related to EGF and PDGF) on adult stem cell destiny are investigated and confirmed.

Interactome analysis which provides useful information about protein complexes and protein–protein interactions can be utilized in mESCs study for solving complexity of development and pluripotency processes of stem cells. Main efforts related to monitoring differentiation, protein–protein interactions controlling self-renewal properties of stem cells were done via a proteomic approach [86–91]. The known role of proteins BAF45a and BAF53a as part of the SWI/SNF and remodeling complex, which are replaced by BAF45b, BAF45c, and BAF53b, as progenitors of the cell cycle in neural stem cells is validated by interactome investigation [92].

4.4 *Proteomic Study of iPSC*

The third type of stem cells that have significant potential for clinical research like disease modeling are known as human induced pluripotent stem cells (iPSCs). These types of stem cells are generated by retroviral transduction. More proteomics, genomics, and metabolomics investigation are required to understand molecular mechanisms which are involved in fate of these stem cells. A proteomic study characterized about 8000 differentially expressed genes between iPSCs and ESCs [93].

Currently, full characterization of hiPSCs via quantitative proteomics is not available, however, undirected several study in the proteome level of hiPSC from different somatic sources are reported. There are quantitative and qualitative adjustments of hiPSC/hESC proteomes for eight hiPSC lines of various somatic sources which are derived in different laboratories with different methods of reprogramming. Numbers of 99 down-regulated and 424 up-regulated proteins are determined for proteome difference between hiPSC isolated from fibroblasts (SB5-MP1) and fibroblast cells [94].

5 Conclusion

Regenerative medicine as a prominent field in treatment has attracted attention of researchers and physicians. Stem cells play a crucial role in regenerative medicine and high throughout methods such as metabolomics, proteomics, and genomics are powerful and suitable technologies to characterize stem cells. Pluripotency, differentiation, and development of stem cells are the major processes that are tied to control and regulation of protein expression activity. Surely progress in instrumental aspects and knowledge of these "omics" technologies will draw new perspective in treatment of many diseases.

References

1. Kingham E, Oreffo R. Embryonic and induced pluripotent stem cells: understanding, creating and exploiting the nano-nich for regenerative medicine. ACS Nano. 2013;3:1867–81.
2. Shyh-Chang N, et al. The metabolic programming of stem cells. Genes Dev. 2017;31:336–46.

3. McNamara L, et al. Metabolomics: a valuable tool for stem cell monitoring in regenerative medicine. Interface. 2012;9:1713–24.
4. Lorelei D, Harley I. Neural stem cells (NSC) and proteomics. Mol Cell Proteomics. 2016; 15:344–54.
5. Aebersold R, Mann M. Mass spectrometry-based proteomics. Nature. 2003;422:198–207.
6. Vlahou A, Fountoulakis M. Proteomic approaches in the search for disease biomarkers. J Chromatogr B Analyt Technol Biomed Life Sci. 2005;814:11–9.
7. Farahani M, Rezaei-Tavirani M, Zali H, Arefi Oskouie A, Omidi M, Lashay A. Deciphering the transcription factormicroRNA- target gene regulatory network associated with graphene oxide cytotoxicity. Nanotoxicology. 2018;12(9):1–13.
8. Kalantari S, Rutishauser D, Samavat S, Nafar M, Mahmudieh L, Rezaei-Tavirani M, et al. Urinary prognostic biomarkers and classification of IgA nephropathy by high resolution mass spectrometry coupled with liquid chromatography. PloS One. 2013;8:808–30.
9. Atkinson S, Armstrong L. Epigenetics in embryonic stem cells: regulation of pluripotency and differentiation. Cell Tissue Res. 2008;331:23–9.
10. Aslam B, Basit M, Nisar M, Khurshid M, Rasool M. Proteomics: technologies and their applications. J Chromatogr Sci. 2017;2:182–96.
11. Avior Y, Sagi I, Benvenisty N. Pluripotent stem cells in disease modeling and drug discovery. Nat. Rev. Mol Cell Biol. 2016;17:170–82.
12. Birey F, Andersen J, Makinson CD, Islam S, Wei W, Huber N, Fan HC, Metzler KRC, Panagiotakos G, Thom N, et al. Assembly of functionally integrated human forebrain spheroids. Nature. 2017;545:54–9.
13. Boroviak T, Loos R, Lombard P, Okahara J, Behr R, Sasaki E, Nichols J, Smith A, Bertone P. Lineage-specific profiling delineates the emergence and progression of naïve pluripotency in mammalian embryogenesis. Dev Cell. 2015;35:366–82.
14. Carbognin E, Betto RM, Soriano ME, Smith AG, Martello G. Stat3 promotes mitochondrial transcription and oxidative respiration during maintenance and induction of naïve pluripotency. EMBO J. 2016;35:618–34.
15. Carey BW, Finley LWS, Cross JR, Allis CD, Thompson CB. Intracellular α-ketoglutarate maintains the pluripotency of embryonic stem cells. Nature. 2015;518:413–6.
16. Chen J, Guo L, Zhang L, Wu H, Yang J, Liu H, Wang X, Hu X, Gu T, Zhou Z, et al. Vitamin C modulates TET1 function during somatic cell reprogramming. Nat Genet. 2013;12:1504–9.
17. Davidson K, Mason EA, Pera MF. The pluripotent state in mouse and human. Development. 2015;142:3090–9.
18. Fukawa T, Yan-Jiang BC, Min-Wen JC, Jun-Hao ET, Huang D, Qian CN, Ong P, Li Z, Chen S, Mak SY, et al. Excessive fatty acid oxidation induces muscle atrophy in cancer cachexia. Nat Med. 2016;22:666–71.
19. Gan B, Hu J, Jiang S, Liu Y, Sahin E, Zhuang L, Fletcher-Sananikone E, Colla S, Wang YA, Chin L, et al. Lkb1 regulates quiescence and metabolic homeostasis of hematopoietic stem cells. Nature. 2010;468:701–4.
20. Goh LH, Zhou X, Lee MC, Lin S, Wang H, Luo Y, Yang X. Clueless regulates aPKC activity and promotes self-renewal cell fate in Drosophila lgl mutant larval brains. Dev Biol. 2013;381:353–64.
21. Gu W, Gaeta X, Sahakyan A, Chan AB, Hong CS, Kim R, Braas D, Plath K, Lowry WE, Christofk HR. Glycolytic metabolism plays a functional role in regulating human pluripotent stem cell state. Cell Stem Cell. 2016;19:476–90.
22. Gurumurthy S, Xie SZ, Alagesan B, Kim J, Yusuf RZ, Saez B, Tzatsos A, Ozsolak F, Milos P, Ferrari F, et al. The Lkb1 metabolic sensor maintains haematopoietic stem cell survival. Nature. 2010;468:659–63.
23. Hochmuth C, Biteau B, Bohmann D, Jasper H. Redox regulation by Keap1 and Nrf2 controls intestinal stem cell proliferation in Drosophila. Cell Stem Cell. 2011;8:188–99.
24. Holmström KM, Finkel T. Cellular mechanisms and physiological consequences of redox-dependent signaling. Nat Rev Mol Cell Biol. 2014;15:411–21.

25. Ito K, Turcotte R, Cui J, Zimmerman SE, Pinho S, Mizoguchi T, Arai F, Runnels JM, Alt C, Teruya-Feldstein J, et al. Self-renewal of a purified Tie2+ hematopoietic stem cell population relies on mitochondrial clearance. Science. 2016;354:1156–60.
26. Karigane D, Kobayashi H, Morikawa T, Ootomo Y, Sakai M, Nagamatsu G, Kubota Y, Goda N, Matsumoto M, Nishimura EK, et al. p38α activates purine metabolism to initiate hematopoietic stem/progenitor cell cycling in response to stress. Cell Stem Cell. 2016;19:192–204.
27. Khaw SL, Min-Wen C, Koh CG, Lim B, Shyh-Chang N. Oocyte factors suppress mitochondrial polynucleotide phosphorylaseto remodel the metabolome and enhance reprogramming. Cell Rep. 2015;12:1080–8.
28. Prieto J, León XP, Sendra R, Bort R, Ferrer-Lorente R, Raya A, López-García C, Torres J. Early ERK1/2 activation promotes DRP1-dependent mitochondrial fission necessary for cell reprogramming. Nat Commun. 2016;7:11124.
29. Moussaieff A, Rouleau M, Kitsberg D, CohenM LG, Barasch D, Nemirovski A, Shen-Orr S, Laevsky I, Amit M, et al. Glycolysis-mediated changes in acetyl-CoA and histone acetylation control the early differentiation of embryonic stem cells. Cell Metab. 2015;21:392–402.
30. Maryanovich M, Zaltsman Y, Ruggiero A, Goldman A, Shachnai L, Zaidman SL, Porat Z, Golan K, Lapidot T, Gross A. An MTCH2 pathway repressing mitochondria metabolism regulates haematopoietic stem cell fate. Nat Commun. 2015;6:7901.
31. Luchsinger LL, de Almeida MJ, Corrigan DJ, Mumau M, Snoeck HW. Mitofusin 2 maintains haematopoietic stem cells with extensive lymphoid potential. Nature. 2016;529:528–5231.
32. Rimmelé P, Liang R, Bigarella CL, Kocabas F, Xie J, Serasinghe MN, Chipuk J, Sadek H, Zhang CC, Ghaffari S. Mitochondrial metabolism in hematopoietic stem cells requires functional FOXO3. EMBO Rep. 2015;16:1164–76.
33. Ryall JG, Dell'Orso S, Derfoul A, Juan A, Zare H, Feng X, Clermont D, Koulnis M, Gutierrez-Cruz G, Fulco M, et al. The NAD+-dependent SIRT deacetylase translates a metabolic switch into regulatory epigenetics in skeletal muscle stem cells. Cell Stem Cell. 2015;16:171–83.
34. Shaltouki A, Sivapatham R, Pei Y, Gerencser AA, Momčilović O, Rao MS, Zeng X. Mitochondrial alterations by PARKIN in dopaminergic neurons using PARK2 patient-specific and PARK2 knockout isogenic iPSC lines. Stem Cell Reports. 2015;4:847–59.
35. Shiraki N, Shiraki Y, Tsuyama T, Obata F, Miura M, Nagae G, Aburatani H, Kume K, Endo F, Kume S. Methionine metabolism regulates maintenance and differentiation of human pluripotent stem cells. Cell Metab. 2014;19:780–94.
36. Shyh-Chang N, Daley GQ. Metabolic switches linked to pluripotency and embryonic stem cell differentiation. Cell Metab. 2015;21:349–50.
37. Spencer JA, Ferraro F, Roussakis E, Klein A, Wu J, Runnels JM, Zaher W, Mortensen LJ, Alt C, Turcotte R, et al. Direct measurement of local oxygen concentration in the bone marrow of live animals. Nature. 2014;508:269–73.
38. Sperber H, Mathieu J, Wang Y, Ferreccio A, Hesson J, Xu Z, Fischer KA, Devi A, Detraux D, Gu H, et al. The metabolome regulates the epigenetic landscape during naive-toprimed human embryonic stem cell transition. Nat Cell Biol. 2015;17:1523–35.
39. Stevens DA, Lee Y, Kang HC, Lee BD, Lee YI, Bower A, Jiang H, Kang SU, Andrabi SA, Dawson VL, et al. Parkin loss leads to PARIS-dependent declines in mitochondrial mass and respiration. Proc Natl Acad Sci. 2015;112:11696–701.
40. Stoll EA, Makin R, Sweet IR, Trevelyan AJ, Miwa S, Horner PJ, Turnbull DM. Neural stem cells in the adult subventricular zone oxidize fatty acids to produce energy and support neurogenic activity. Stem Cells. 2015;33:2306–19.
41. Ware CB, Nelson AM, Mecham B, et al. Derivation of naïve human embryonic stem cells. Proc Natl Acad Sci. 2014;111:4484–9.
42. Vannini N, Girotra M, Naveiras O, Nikitin G, Campos V, Giger S, Roch A, Auwerx J, Lutolf MP. Specification of haematopoietic stem cell fate via modulation of mitochondrial activity. Nat Commun. 2016;7:13125.
43. Ueda T, Nagamachi A, Takubo K, Yamasaki N, Matsui H, Kanai A, Nakata Y, Ikeda K, Konuma T, Oda H, et al. Fbxl10 overexpression in murine hematopoietic stem cells induces leukemia involving metabolic activation and upregulation of Nsg2. Blood. 2015;125:3437–46.

44. Tan JL, Fogley RD, Flynn RA, Ablain J, Yang S, Saint-André V, Fan ZP, Do BT, Laga AC, Fujinaga K, et al. Stress from nucleotide depletion activates the transcriptional regulator HEXIM1 to suppress melanoma. Mol Cell. 2016;62:34–46.
45. TeSlaa T, Chaikovsky AC, Lipchina I, Escobar SL, Hochedlinger K, Huang J, Graeber TG, Braas D, Teitell MA. α-Ketoglutarate accelerates the initial differentiation of primed human pluripotent stem cells. Cell Metab. 2016;24:485–93.
46. Takashima Y, Guo G, Loos R, Nichols J, Ficz G, Krueger F, Oxley D, Santos F, Clarke J, Mansfield W, et al. Resetting transcription factor control circuitry toward ground-state pluripotency in human. Cell. 2014;158:1254–69.
47. Xie Z, Jones A, Deeney JT, Hur SK, Bankaitis VA. Inborn errors of long-chain fatty acid β-oxidation link neural stem cell self-renewal to autism. Cell Rep. 2016;14:991–9.
48. Zhang H, Badur MG, Divakaruni AS, Parker SJ, Jäger C, Hiller K, Murphy AN, Metallo CM. Distinct metabolic states can support self-renewal and lipogenesis in human pluripotent stem cells under different culture conditions. Cell Rep. 2016;16:1536–47.
49. Gore A, Li Z, Fung H-L, Young J, Agarwal S, et al. Somatic coding mutations in human induced pluripotent stem cells. Nature. 2011;471:63–7.
50. Hussein SM, Batada NN, Vuoristo S, Ching RW, Autio R, et al. Copy number variation and selection during reprogramming to pluripotency. Nature. 2011;471:58–62.
51. Laurent LC, Ulitsky I, Slavin I, Tran H, Schork A, et al. Dynamic changes in the copy number of pluripotency and cell proliferation genes in human ESCs and iPSCs during reprogramming and time in culture. Cell Stem Cell. 2011;8:106–18.
52. Quinlan AR, Boland MJ, Leibowitz ML, Shumilina S, Pehrson SM, et al. Genome sequencing of mouse induced pluripotent stem cells reveals retroelement stability and infrequent DNA rearrangement during reprogramming. Cell Stem Cell. 2011;9:366–73.
53. Riggs JW, Barrilleaux BL, Chan V, Knoepfler PS. Induced pluripotency and oncogenic transformation are distinct, but highly related processes. Stem Cells Dev. 2012;22(1):37–50.
54. Barupal DK, Haldiya PK, Wohlgemuth G, Kind T, Kothari SL, et al. MetaMapp: mapping and visualizing metabolomic data by integrating information from biochemical pathways and chemical and mass spectral similarity. BMC Bioinform. 2012;13:99.
55. Bhutani N, Brady JJ, Damian M, Sacco A, Corbel SY, Blau HM. Reprogramming towards pluripotency requires AID-dependent DNA demethylation. Nature. 2010;463:1042–7.
56. Chambers I, Silva J, Colby D, et al. Nanog safeguards pluripotency and mediates germline development. Nature. 2007;450:1230–4.
57. Judson RL, Babiarz JE, Venere M, Blelloch R. Embryonic stem cell-specific microRNAs promote induced pluripotency. Nat Biotechnol. 2009;27:459–61.
58. Jia F, Wilson KD, Sun N, Gupta DM, Huang M, Li Z, Panetta NJ, Chen ZY, Robbins RC, Kay MA, Longaker MT, Wu JC. A nonviral minicircle vector for deriving human iPS cells. Nat Methods. 2010;7:197–9.
59. Itzhaki I, Maizels L, Huber I, Zwi-Dantsis L, Caspi O, Winterstern A, Feldman O, Gepstein A, Arbel G, Hammerman H, Boulos M, Gepstein L. Modeling the long QT syndrome with induced pluripotent stem cells. Nature. 2011;471:225–9.
60. Ramachandra CJ, Shahbazi M, Kwang TW, Choudhury Y, Bak XY, Yang J, Wang S. Efficient recombinase-mediated cassette exchange at the AAVS1 locus in human embryonic stem cells using baculoviral vectors. Nucleic Acids Res. 2011;39:107.
61. Allende ML, Cook EK, Larman BC, Nugent A, Brady JM, Golebiowski D, Sena-Esteves M, Tifft CJ, Proia RL. Cerebral organoids derived from Sandhoff disease-induced pluripotent stem cells exhibit impaired neurodifferentiation. J Lipid Res. 2018;59:550–63.
62. Ardhanareeswaran K, Mariani J, Coppola G, Abyzov A, Vaccarino M. Human induced pluripotent stem cells for modeling neurodevelopmental disorders. Nat Rev Neurol. 2017;13:265–78.
63. Issaq H, Veenstra T. Two-dimensional polyacrylamide gel electrophoresis (2D-PAGE): advances and perspectives. Biotechniques. 2008;44:697–700.
64. Marouga R, David S, Hawkins E. The development of the DIGE system: 2D fluorescence difference gel analysis technology. Anal Bioanal Chem. 2005;382:669–78.

65. Kroksveen AC, Jaffe JD, Aasebo E, Barsnes H, Bjorlykke Y. Quantitative proteomics suggests a decrease in the secretogranin-1 cerebrospinal fluid levels during the disease course of multiple sclerosis. Proteomics. 2015;15:3361–9.
66. Cravatt BF, Simon GM, Yates JR. The biological impact of mass-spectrometry-based proteomics. Nature. 2007;450:991–1000.
67. Dai B, Rasmussen TP. Global epiproteomic signatures distinguish embryonic stem cells from differentiated cells. Stem Cells. 2007;25:2567–74.
68. Freitas MA, Sklenar AR, Parthun MR. Application of mass spectrometry to the identification and quantification of histone post-translational modifications. J Cell Biochem. 2004;92: 691–700.
69. Ginis I, Luo Y, Miura T, Thies S, Brandenberger R, Gerecht-Nir S, Amit M, Hoke A, Carpenter MK, Itskovitz-Eldor J, Rao MS. Differences between human and mouse embryonic stem cells. Dev Biol. 2004;269:360–80.
70. Ivanova N, Dobrin R, Lu R, Kotenko I, Levorse J, DeCoste C, Schafer X, Lun Y, Lemischka IR. Dissecting self-renewal in stem cells with RNA interference. Nature. 2006;442:533–8.
71. Josephson R, Ording CJ, Liu Y, Shin S, Lakshmipathy U, Toumadje A, Love B, Chesnut JD, Andrews PW, Rao MS, Auerbach JM. Qualification of embryonal carcinoma 2102Ep as a reference for human embryonic stem cell research. Stem Cells. 2007;25:437–46.
72. Kirkpatrick DS, Denison C, Gygi SP. Weighing in on ubiquitin: the expanding role of mass-spectrometry-based proteomics. Nat Cell Biol. 2005;7:750–7.
73. Hassani SN, Totonchi M, Gourabi H. Signaling roadmap modulating naive and primed pluripotency. Stem Cells Dev. 2014;23:193–208.
74. Hughes C, Radan L, Wing Y, William L, Dean H, Gilles A. Mass spectrometry-based proteomic analysis of the matrix microenvironment in pluripotent stem cell culture. Mol Cell Proteomics. 2012;11:1924–36.
75. Shekari F, Han CL, Lee J, Mirzaei M, Gupta V, Haynes PA, Lee B, Baharvand H, Chen YJ, Hosseini Salekdeh G. Surface markers of human embryonic stem cells: a meta-analysis of membrane proteomics reports. Expert Rev Proteomics. 2018;55:1–21.
76. Stanton LW, Bakre MM. Genomic and proteomic characterization of embryonic stem cells. Curr Opin Chem Biol. 2007;11:399–404.
77. Sze SK, de Kleijn DP, Lai RC, Khia Way Tan E, Zhao H, Yeo KS, Low TY, Lian Q, Lee CN, Mitchell W, El Oakley RM, Lim SK. Elucidating the secretion proteome of human ESC derived mesenchymal stem cells. Mol Cell Proteomics. 2007;6:1680–9.
78. Van Hoof D, Passier R, Ward-Van Oostwaard D, Pinkse MW, Heck AJ, Mummery CL, Krijgsveld J. A quest for human and mouse embryonic stem cell-specific proteins. Mol Cell Proteomics. 2006;5:1261–73.
79. Van Hoof D, Mummery CL, Heck AJ, Krijgsveld J. Embryonic stem cell proteomics. Proteomics. 2006;3:427–37.
80. Van Hoof D, Pinkse MW, Oostwaard DW, Mummery CL, Heck AJ, Krijgsveld J. An experimental correction for arginine-to-proline conversion artifacts in SILACbased quantitative proteomics. Nat Methods. 2007;4:677–8.
81. Wu Q, Chen X, Zhang J, Loh YH, Low TY, Zhang W, Zhang W, Sze SK, Lim B, Ng HH. Sall4 interacts with Nanog and co-occupies Nanog genomic sites in embryonic stem cells. J Biol Chem. 2006;281:24090–4.
82. Zhao S, Nichols J, Smith AG, Li M. SoxB transcription factors specify neuroectodermal lineage choice in ES cells. Mol Cell Neurosci. 2004;27:332–42.
83. Bryder DR, Weissman DJ, IL. Hematopoietic stem cells: the paradigmatic tissue-specific stem cell. Am J Pathol. 2006;169:338–46.
84. Kern S, Eichler H, Stove J, Kluter H, Bieback K. Comparative analysis of mesenchymal stem cells from bone marrow, umbilical cord blood, or adipose tissue. Stem Cells. 2006;24:1294–301.
85. Pevsner-Fischer M, Levin S, Zipori D. The origins of mesenchymal stromal cell heterogeneity. Stem Cell Rev. 2011;7:560–8.

86. Hematti P. Human embryonic stem cell-derived mesenchymal stromal cells. Transfusion. 2011;51:138S–44S.
87. de Peppo GM, et al. Human embryonic mesodermal progenitors highly resemble human mesenchymal stem cells and display high potential for tissue engineering applications. Tissue Eng Part A. 2010;16:2161–82.
88. Vodyanik MA, et al. A mesoderm-derived precursor for mesenchymal stem and endothelial cells. Cell Stem Cell. 2010;7:718–29.
89. Kimbrel EA, et al. Mesenchymal stem cell population derived from human pluripotent stem cells displays potent immunomodulatory and therapeutic properties. Stem Cells Dev. 2014;23:1611–24.
90. Sanchez L, et al. Enrichment of human ESC-derived multipotent mesenchymal stem cells with immunosuppressive and anti- inflammatory properties capable to protect against experimental inflammatory bowel disease. Stem Cells. 2011;29:251–62.
91. Wang X, et al. Human ESC-derived MSCs outperform bone marrow MSCs in the treatment of an EAE model of multiple sclerosis. Stem Cell Reports. 2014;3:115–30.
92. Liberski AR, et al. Adaptation of a commonly used, chemically defined medium for human embryonic stem cells to stable isotope labeling with amino acids in cell culture. J Proteome Res. 2013;12:3233–45.
93. Yamana R, Iwasaki M, Wakabayashi M, Nakagawa M, Yamanaka S, Ishihama Y. Rapid and deep profiling of human induced pluripotent stem cell proteome by one-shot NanoLC–MS/MS analysis with meter-scale monolithic silica columns. J Proteome Res. 2012;44:1–12.
94. Pripuzova N, Getie-Kebtie M, Grunseich C, Sweeneyc C, Malechc H, Alterman M. Development of a protein marker panel for characterization of human induced pluripotent stem cells (hiPSCs) using global quantitative proteome analysis. Stem Cell Res. 2015;14:323–38.

Metabolic Profiling of the Mesenchymal Stem Cells' Secretome

Kambiz Gilany, Mohammad Javad Masroor, Arash Minai-Tehrani, Ahmad Mani-Varnosfaderani, and Babak Arjmand

Abstract Human 'multipotent mesenchymal stromal cells' (hMSCs) are popular cells in human regenerative medicine due to their ability to renew themselves and differentiate into various specialized cell types under specific physiological or experimental conditions. HMSCs secrete a broad spectrum of components including proteins and metabolites that represent significant effects on the cells in their neighborhood. Furthermore, it assists to characterize them. The therapeutic effects of hMSCs were thought to be due to their multipotent characterization and their ability to engraft and differentiate at the site of injury. However, *recent studies* have *revealed* the fact that a secretome plays important role in the therapeutic potential of hMSCs. Here, we outline the decoding of the metabolome for hMSCs secretome through metabolic profiling by application of MALDI-TOF-MS which could contribute to a better understanding of the therapeutic effects associated with hMSCs and their characterization.

Keywords Metabolomics · Secretome · Metabolites · Mesenchymal stem cells · MALDI-TOF-MS

K. Gilany
Department of Biomedical Sciences, University of Antwerp, Antwerp, Belgium

Reproductive Biotechnology Research Center, Avicenna Research Institute, ACECR, Tehran, Iran

M. J. Masroor · A. Mani-Varnosfaderani
Chemometrics Laboratory, Department of Chemistry, Tarbiat Modares University, Tehran, Iran

A. Minai-Tehrani
Nanobiotechnology Research Center, Avicenna Research Institute, ACECR, Tehran, Iran

B. Arjmand (✉)
Cell Therapy and Regenerative Medicine Research Center, Endocrinology and Metabolism Molecular-Cellular Sciences Institute, Tehran University of Medical Sciences, Tehran, Iran

Metabolomics and Genomics Research Center, Endocrinology and Metabolism Molecular-Cellular Sciences Institute, Tehran University of Medical Sciences, Tehran, Iran
e-mail: barjmand@sina.tums.ac.ir

B. Arjmand (ed.), *Genomics, Proteomics, and Metabolomics*, Stem Cell Biology and Regenerative Medicine, https://doi.org/10.1007/978-3-030-27727-7_3

1 Introduction

Human 'multipotent mesenchymal stromal cells' (hMSCs) are one of the most encouraging types of stem cells for cell-based therapies. An increased clinical application of hMSCs during the last three decades is apparently observed, as shown by the growing number of studies. The therapeutic effects of hMSCs in various diseases suggested to be caused by their capacity to self-renew, differentiate into different cell types, to secrete product (secretome) factors with paracrine actions, be immunosuppressive and finally have immunomodulatory properties [1–5]. Furthermore, hMSCs have attracted great interest because of their ability of expansion in vitro and isolation from a wide range of ethically accepted adults [6, 7].

Russian scientist Alexander Friedenstein (1924–1998) and coworkers were the first to isolate and describe hMSCs found from bone marrow in a series of studies in the early 1960s and 1970s. They showed that hMSCs can be differentiated into osteocytes, chondrocytes, adipocytes, and myoblasts [8–10]. Nowadays, hMSCs can be isolated from almost all tissues including perivascular area [11]. Inspite of the presence of hMSCs in different tissues and their similar characteristics, their molecular and maybe phenotype features reflect their tissue of origin [3, 12].

In 2006, because the definition of hMSCs was inconsistent among investigators, the International Society for Cell Therapy (ISCT) agreed a set of criteria to define hMSCs. First, hMSCs must show the ability to adhere to the plastic when maintained in standard culture conditions using tissue culture flasks. Second, they must express specific surface markers ($\geq$95%) such as CD105, CD73, and CD90, as measured by flow cytometry, whereas these cells must lack the expression ($\leq$2% positive) of CD45, CD34, CD14 or CD11b, CD79a or CD19, and HLA class II. Third, the cells must be able to undergo tri-lineage differentiation including osteoblasts, adipocytes, and chondroblasts under standard in vitro differentiating conditions [3].

Beside the special property of multipotency, hMSCs have several unique characterizations, such as immunomodulatory features; secrete cytokines and immune receptors which regulate the microenvironment in the host tissue. This stem cell has been clinically tested for treatment of wide range of pathological conditions such as rheumatoid arthritis, Crohn's disease, graft versus host disease (GVHD) in bone marrow transplantation, type I diabetes, myocardial infraction, cardiovascular diseases, and amyotrophic lateral sclerosis [11, 13, 14].

The mechanism of hMSCs therapeutic effects is under debate. It is suggested that hMSCs attribute to migration to the sites of injury and inflammation, engraft into the damaged tissues and differentiate into specialized cell types. However, recent studies have discovered that implanted hMSCs did not survive for long, and therapeutic activity of hMSCs could be due to the secreted product (secretome) including proteins and metabolites [15–17]. Three main mechanisms have been proposed behind the therapeutic effects of hMSCs. Firstly, its homing ability, for which systemically transplanted hMSCs migrate to the focus of acute injury due to chemical gradients. It is hypothesized that this mechanism is like a process similar to that of leukocyte migration. Secondly, hMSCs differentiation into multiple cell types, which locally engraft and induce restoration of function by augmenting or replacing damaged tissues. Finally, the hMSCs secrete bioactive factors, which may potentially affect both local and systemic physiological processes [15]. Although, several studies support these hypothesized actions, the mechanism of hMSCs action is not completely clear. Recently, the stem cell community is trying to understand the mechanism of hMSCs action through its properties of secretome with autocrine and paracrine actions and uncover hMSCs potential in cell therapy and regenerative medicine.

In context to cell therapy and regenerative medicine, metabolomics is a state of the art of analytical technique for biomaterials analysis. Metabolomics shows the comprehensive and systematic analysis of biological systems. Metabolomics has several advantages compared to proteomics; it can detect a large number of metabolites with very small quantities of sample material. Furthermore, in metabolomics, sample preparation is less demanding than proteomics [18, 19].

The studies of secreted products (metabolites) of hMSCs through metabolomics may contribute to a better understanding of hMSCs therapeutic effects.

2 Metabolomics of Human Stem Cells

The autocrine and paracrine effects of hMSCs are due to the synthesis and secretion of the broad spectrum of bioactive components including proteins and metabolites. In this study, we only focus on the metabolites which are identified and secreted by hMSCs.

Metabolites are the final products of cellular regulatory processes, and their levels can be accepted as an end response of biological systems to genetic or environmental alterations. In similar to the term 'transcriptome' and 'proteome', the expressed metabolites by the biological systems compose its 'metabolome' [20]. The term metabolomics was coined at the end of the 1990s to describe the progress of techniques which the purpose was to identify and quantify all the metabolites that are present within a cell, tissue, or organism during a genetic modification or physiological stimulus, Fig. 1 [21, 22]. Nowadays, metabolomics has developed to the state of the art 'omics' technologies and different terms are defined based on the applied approach [19]. Today, human metabolome database contains more than 114,000 metabolites [23]. However, metabolomics in the stem cell research is in early stage [24, 25].

To the best of our knowledge, no mapping of metabolome of hMSCs secreted metabolites has been done yet. However, a shift toward metabolomics by stem cell community have published a handful of metabolites identified from stem cells [15, 26–31]. Table 1 shows the identified metabolites from stem cells by different metabolomics techniques.

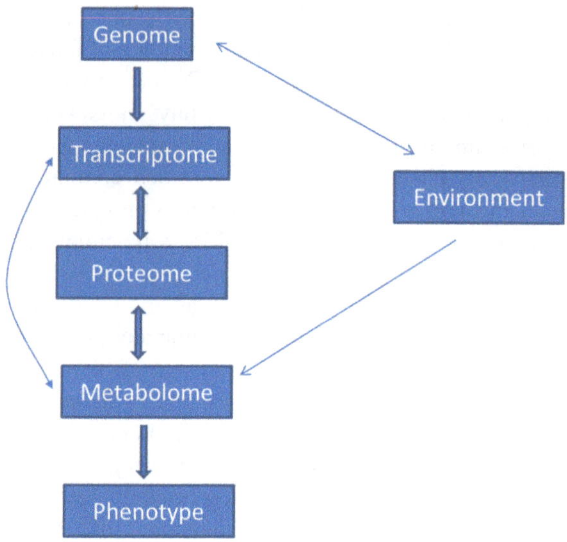

Fig. 1 A schematic view of metabolomics. Metabolomics position is shown with respect to the other omics and the stimulus of the environment

As shown in the Table 1, only four types of stem cells have been used for metabolomics studies so far. These are human embryonic stem cell (hES), human bone marrow mesenchymal stem cells (hBMMSCs), human adipose-derived stem cells (hASCs), and finally hMSCs.

Figure 2 shows a Venn diagram of the identified metabolites in the stem cells. Figure 2 shows that there are only nine overlapping metabolites. Most importantly it shows that metabolomics analysis of stem cells can be used for characterization of human stem cells. As example, the overlap between hASCs and hMSCs is zero.

Furthermore, Venn diagram figure shows that the techniques are complementary. However, it is well-established that mass spectrometry (MS) based metabolomics has the power to identify much more metabolites than other techniques. The drawback of MS-based metabolomics demonstrates that all peaks can not be assigned to one specific metabolite [33, 34].

The biological molecules synthesized in the development of the proliferation and differentiation of stem cells could be secreted to the extracellular environment. Cell culturing in vitro could begin changes in the amount of particular bioactive factors secreted in the culture media. The secreted molecules could act as a possible replacement of stem cells for therapeutic uses to overcome several stem cell interrelated issues including cell origin and immunocompatibility. The therapeutic use of stem cell secretome could show new, safe, and effective approaches with anticipated outcomes as alternative to the cell therapy [35–38].

Table 1 Metabolites identified from stem cell

Stem cell	Metabolite (HMDB ID)	Technique	Reference
hES; hMSCs	Kynurenine	LCMS	[31, 32]
hES; hMSCs; hASCs	Glutamate	1D-NMR; LCMS	[27, 28, 30, 31]
hES	Pyroglutamic acid	LCMS	[31]
hES	Glutathione	LCMS	[31]
hES	S-adenosyl-homocysteine	LCMS	[31]
hES	Folate	LCMS	[31]
hES	Aminobutyric acid	LCMS	[31]
hMSCs	3-Hydroxyisovalerate	1D-NMR	[30]
hMSCs	ADP	1D-NMR	[30]
hMSCs	ATP	1D-NMR	[30]
hMSCs; hBMMSCs	Acetate	1D-NMR; GCMS	[26, 28, 30]
hMSCs; hBMMSCs	Alanine	1D-NMR; GCMS	[26, 30]
hMSCs; hES; hASCs	Aspartate	1D-NMR; GCMS	[27, 29, 30]
hMSCs	Betaine	1D-NMR	[30]
hMSCs	Choline	1D-NMR	[28, 30]
hMSCs	Citrate	1D-NMR	[30]
hMSCs	Creatine	1D-NMR	[30]
hMSCs	Formate	1D-NMR	[28, 30]
hMSCs; hBMMSCs; hASCs	Glycine	1D-NMR; GCMS	[26, 27, 30]
hMSCs; hES; hBMMSCs; hASCs	Isoleucine	1D-NMR; GCMS	[26–30]
hMSCs; hES; hBMMSCs	Lactate	1D-NMR; GCMS	[26, 28–30]
hMSCs; hES; hBMMSCs; hASCs	Leucine	1D-NMR; GCMS	[26–30]
hMSCs	NAD+	1D-NMR	[30]
hMSCs	O-Phosphocholine	1D-NMR	[30]
hMSCs; hES; hBMMSCs; hASCs	Proline	1D-NMR; GCMS	[26, 27, 29, 30]
hMSCs	Taurine	1D-NMR	[30]
hMSCs; hASCs; hMSCs	Tyrosine	1D-NMR	[27, 28, 30]
hMSCs	UDP-glucose	1D-NMR	[30]
hMSCs; hES; hBMMSCs; hASCs	Valine	1D-NMR, GCMS	[26–30]
hMSCs; hES; hBMMSCs	Myo-inositol	1D-NMR; GCMS	[26, 29, 30]
hMSCs	Sn-glycero-3-phosphocholine	1D-NMR	[30]
hES	Cholesterol	GCMS	[29]
hES; hASCs	Fatty acids	LC/MS, CE/MS, GCMS	[27, 29]
hES; hBMMSCs; hMSCs	Glucose	1D-NMR; GCMS	[26, 28, 29]

(continued)

Table 1 (continued)

Stem cell	Metabolite (HMDB ID)	Technique	Reference
hES; hBMMSCs; hASCs; hMSCs	Glutamine	1D-NMR; GCMS	[26–29]
hES	Glycerol	GCMS	[29]
hES	Glycogen	GCMS	[29]
hES; hASCs; hMSCs	Histidine	1D-NMR; GCMS	[27–29]
hES; hBMMSCs; hASCs; hMSCs	Lysine	1D-NMR; GCMS	[26–29]
hES; hASCs	Organic acids	GCMS	[29]
hES; hBMMSCs; hASCs; hMSCs	Phenylalanine	GCMS	[26–29]
hES	Phospholipids	GCMS	[29]
hES	Phosphate	GCMS	[29]
hES; hBMMSCs; hASCs	Serine	GCMS	[26, 27, 29]
hES; hASCs; hMSCs	Threonine	1D-NMR;GCMS	[27–29]
hES; hASCs	Tyrosine	GCMS	[27, 29]
hBMMSCs	Oxalic acid	GCMS	[26]
hBMMSCs	Methylmalonic acid	GCMS	[26]
hBMMSCs	Urea	GCMS	[26]
hBMMSCs	Succinic acid	GCMS	[26]
hBMMSCs	Glyceric acid	GCMS	[26]
hBMMSCs	Treonine	GCMS	[26]
hBMMSCs	Aminomalonic acid	GCMS	[26]
hBMMSCs	Pyroglutamic acid	GCMS	[26]
hBMMSCs	Threonic acid	GCMS	[26]
hBMMSCs; hASCs	Asparagine	GCMS	[26, 27]
hBMMSCs	Lauric acid	GCMS	[26]
hBMMSCs	Citric acid	GCMS	[26]
hBMMSCs	Altrose	GCMS	[26]
hBMMSCs	Fructose	GCMS	[26]
hBMMSCs	Galactose	GCMS	[26]
hBMMSCs	Sorbitol	GCMS	[26]
hBMMSCs	Manitol	GCMS	[26]
hBMMSCs	Palmitic acid	GCMS	[26]
hBMMSCs	Stearic acid	GCMS	[26]
hBMMSCs	Sucrose	GCMS	[26]
hBMMSCs	Lactose	GCMS	[26]
hASCs	Glycerolipids	LC/MS, CE/MS, GCM/MS	[27]
hASCs	Glycerophospholipids	LC/MS, CE/MS, GCM/MS	[27]
hASCs	Sphingolipids	LC/MS, CE/MS, GCM/MS	[27]

(continued)

Table 1 (continued)

Stem cell	Metabolite (HMDB ID)	Technique	Reference
hASCs	Sterol lipids	LC/MS, CE/MS, GCM/MS	[27]
hASCs	Carbohydrates	LC/MS, CE/MS, GCM/MS	[27]
hASCs	Vitamin D3	LC/MS, CE/MS, GCM/MS	[27]
hMSCs	Arginine	1D-NMR	[28]
hMSCs	Methionine	1D-NMR	[28]
hMSCs	Pyruvate	1D-NMR	[28]
hMSCs	Tryptophan	1D-NMR	[28]
hMSCs	Nicotinamide	1D-NMR	[28]

hES Human embryonic stem cell, *hBMMSCs* human bone marrow mesenchymal stem cells, *hASCs* human adipose-derived stem cells, *hMSCs* human multipotent mesenchymal stromal cells, *1D-NMR* one dimensional nuclear magnetic resonance, *LC/MS* liquid chromatography mass spectrometry, *CE/MS* capillary electrophoresis mass spectrometry, *GC/MS* gas chromatography mass spectrometry

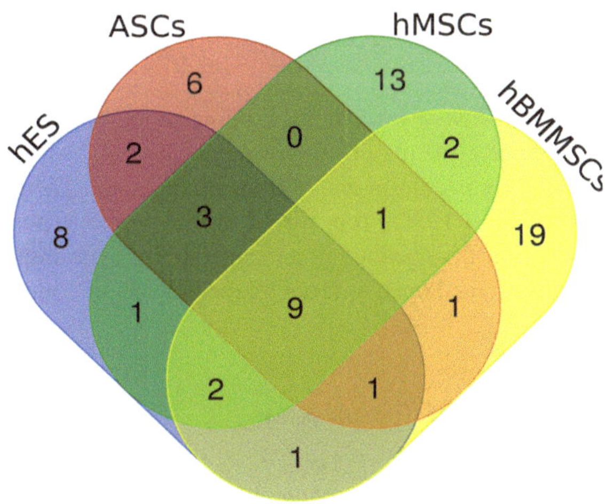

Fig. 2 Venn diagram of the number of unique identified metabolites from human stem cells

3 Metabolic Profiling of MSCs' Secretome by MALDI-TOF-MS

Mass spectrometry technology such as Matrix Assisted Laser Desorption Ionization (MALDI) has been helpful on further analysis of proteins, lipids, and metabolites in the biological samples. MALDI–TOF MS analysis is a sensitive method and it is relatively insensitive to impurities that allows for convenient sample preparation, making it an excellent analytical tool for rapid screening of components in the biological matrices [39]. Additionally, very small amounts of sample are needed for analysis by MALDI-TOF-MS. We propose a metabolomic screening strategy by MALDI-TOF-MS to verify the metabolic profile of culture media. More specifically, in this context, MSCs differentiation in specific biological conditions and also their development, can be studied and monitored through culture media metabolome (secretome).

4 Sample Collection and Preparation

4.1 Cell Culturing

Human BM-MSCs were used in this study. Standard cell culture procedures were used. Human BM-MSCs were characterized including confirmation of ability to undergo adipogenesis, osteogenesis, and chondrogenesis.

4.2 Metabolite Extraction

Metabolites were extracted according to the protocol proposed by Bligh and Dyer [40]. Briefly, 40 μL of culture media was placed in a microcentrifuge tube with 200 μL of chloroform and 100 μL of methanol. The mixture was submitted to centrifugation at 13,000 rpm for 15 min. A biphasic phase was observed. The water- and organic phase containing the metabolites was transferred to another tube, which was maintained open at room temperature until complete evaporation of solvents.

4.3 MALDI Techniques

The air-dried extracted metabolome was mixed with 7 mg/mL DHB dissolved in 0.01% TFA in methanol 70% (v/v) and 1 μL of this solution was spotted onto a stainless MALDI target, dried at room temperature. MS spectra were obtained using an applied biosystems 4800 MALDI TOF/TOF mass spectrometer. An accelerating voltage of 78 mV (sample plate) was used, and the laser operated at a frequency of 1000 Hz. The spectra were recorded in reflectron mode within a mass range of m/z 50–900 in the positive ionization mode. The samples were ionized by 800 shots from the MALDI laser at random spots.

4.4 Bioinformatics

Mass spectra data obtained by MALDI-TOF-MS was processed by MATLAB. Data sets were subjected to multivariate statistical analysis to identify similarities and differences among samples. In order to achieve a good predictive performance, input data matrix must be auto-scaling properly. Principal component analysis (PCA) as an unsupervised statistical method for finding the structure of the data was used in this study, PCA models are shown as score plots and contain of tri-lineage differentiations of BM-MSCs after 4 weeks: principal component (PC) 1 and 2. These plots show groups of samples based on spectral variation.

4.5 Applications of Metabolomics in MSCs

There are ongoing projects to apply metabolomics to monitor stem cell research program. NMR spectroscopy has been used for metabolomics analysis of hES, human umbilicard cord blood plasma (hUCBP), and MSCs' unconditioned and conditioned medium. A wide range of metabolites have been reported from these studies including several amino acids, organic acid derivatives (lactate, formate, and creatine), glucose, and nucleotides (adenosine, uridine, and guanosine species) [28]. Most recently, a NMR metabolomics study of stem cell has been used for the discrimination analysis [41].

MALDI-TOF-MS has a wide application including lipid analysis [39]. To the best of our knowledge, no study has used MALDI-TOF-MS for the study of stem cells metabolome, specifically MSCs secretome.

4.6 Metabolomic Analysis of MSCs' Secretome in Differentiation State Using MALDI-TOF-MS

Metabolomics is an emerging field; however, it has a limited application in stem cell research. One of the challenges of stem cell research is characterization of stem cells [3]. Metabolomics analysis of secretome has the potential to characterize the stem cells. Figure 3 shows MALDI-TOF-MS spectrum of metabolic fingerprinting of secretome from BM-MSCs before differentiation and 1 week after tri-lineage differentiation to osteoblasts, adipocytes, and chondroblasts under standard in vitro differentiating conditions.

Figure 3 shows metabolic fingerprinting technique can be used to characterize the stem cells as soon as 1 week of differentiation.

Furthermore, tri-lineage differentiations of MSCs in 4 weeks were followed up by MALDI-TOF-MS metabolic fingerprinting technique. Figure 4 shows the

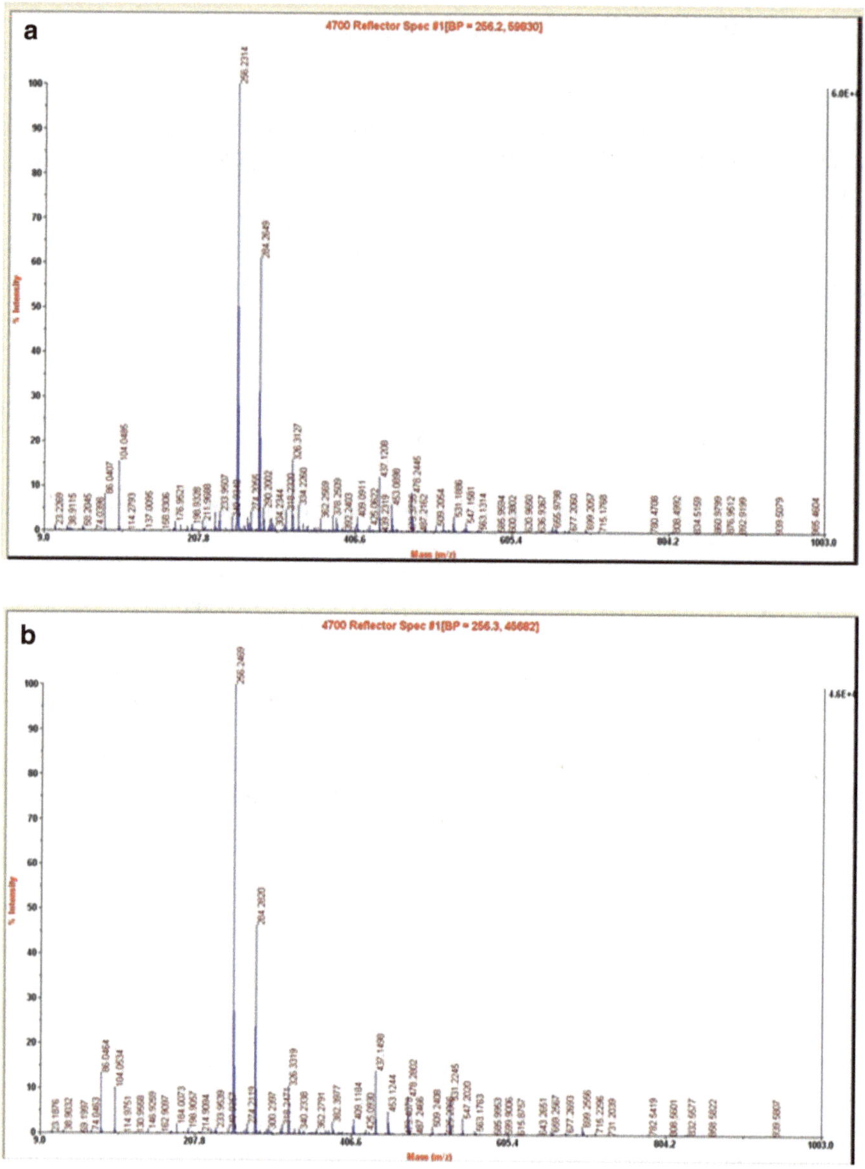

Fig. 3 MALDI-TOF-MS spectrum from BM-MSCs (**a**) undifferentiation state (**b**) differentiation to osteoblasts, (**c**) differentiation to adipocytes, and (**d**) differentiation to chondroblasts under standard in vitro differentiating conditions

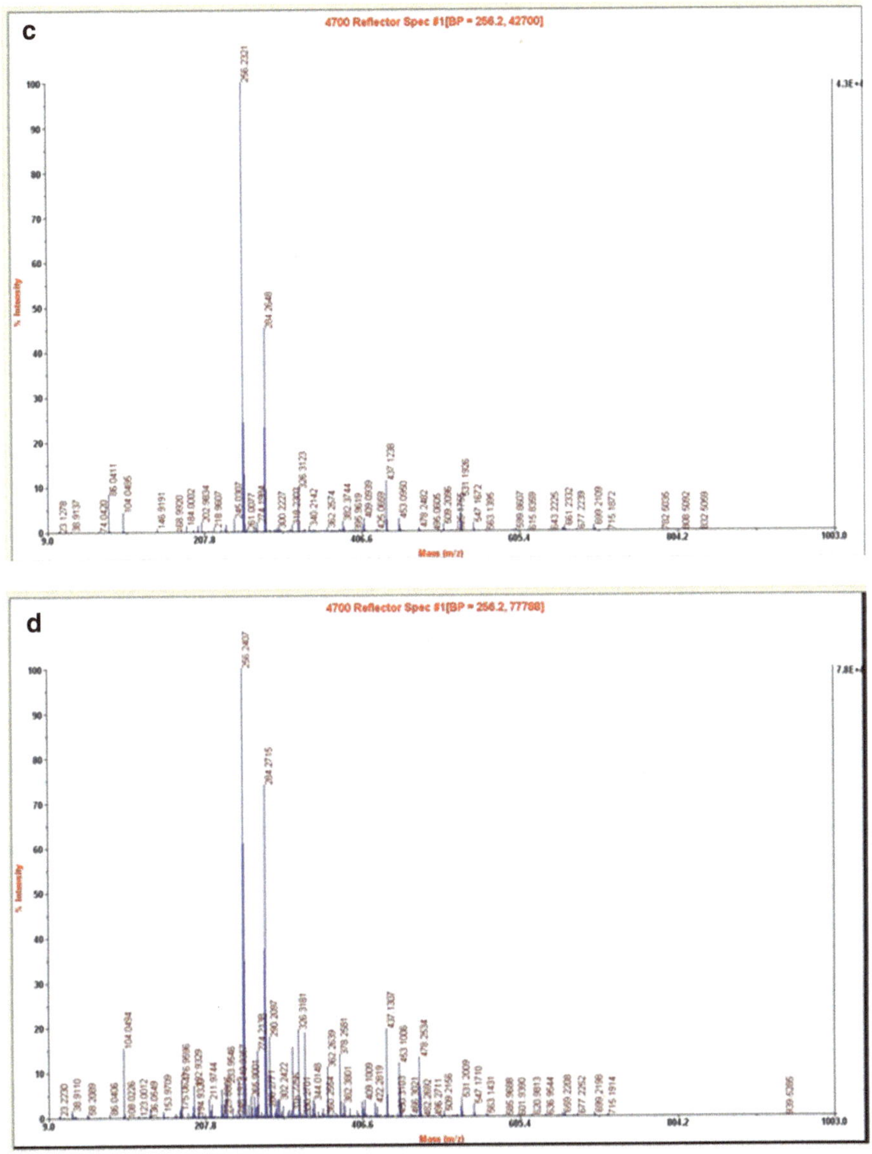

Fig. 3 (continued)

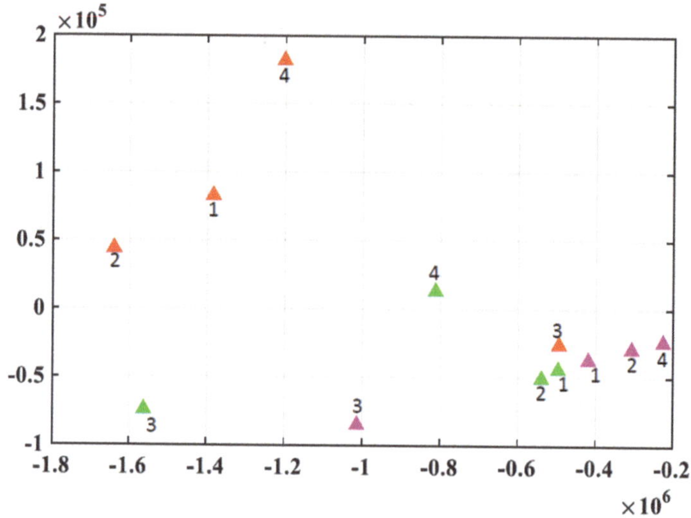

Fig. 4 PCA of tri-lineage differentiations of BM-MSCs after 4 weeks. Weeks are denoting by 1, 2, 3, and 4. Green triangle are BM-MSCs derived osteogenic, red triangle are BM-MSCs derived chondrogenic, and violet triangle are BM-MSCs derived adipogenic

principal component analysis (PCA) of differentiations of tri-lineages under standard in vitro conditions.

The Fig. 4 shows alterations of BM-MSCs secretome during their differentiation. Additionally, Fig. 4 illustrates the fully differentiated state of the secretome of BM-MSCs tri-lineage at different conditions. This means that the stem cells secretome at fully differentiated state has a unique characterization.

Metabolic fingerprinting has been shown to be a powerful tool for diagnosis [3, 42–45]. We have applied metabolic fingerprinting of the secretome for characterization of MSCs. Mass spectrometry-based metabolic fingerprinting has recently developed for detection of different cancer types [46].

Although metabolomics is a fast moving omics field, some main challenges remained to be solved. The most significant issue in metabolomics is bioinformatics part. This metabolomics niche is under development [47]. However, we were not able to identify logical metabolites from our m/z peaks from MALDI-TOF-MS spectrum using metabolome database such as METLIN or HMDB database [23, 48]. The most intense MALD-TO-MS m/z peaks (256.2 and 284.2) did not show any hit.

5 Conclusion

In brief, metabolomics studies so far have shown the richness of information available for stem cell research. Although, the metabolomics in stem cell research is still in its infancy, there are almost 100 identified metabolites that were extracted from different source of stem cells (Table 1). In our opinion, the knowledge gained from metabolome studies of stem cells can be used for cell therapy and regenerative medicine.

Furthermore, our studies indicate that the metabolic fingerprinting of secretome from MSCs can be considered as a fast and simple screening methodology for characterization of stem cell under differentiation state.

References

1. Mendicino M, Bailey AM, Wonnacott K, Puri RK, Bauer SR. MSC-based product characterization for clinical trials: an FDA perspective. Cell Stem Cell. 2014;14(2):141–5.
2. Bianco P, Robey PG, Simmons PJ. Mesenchymal stem cells: revisiting history, concepts, and assays. Cell Stem Cell. 2008;2(4):313–9.
3. Dominici M, Le Blanc K, Mueller I, Slaper-Cortenbach I, Marini F, Krause DS, et al. Minimal criteria for defining multipotent mesenchymal stromal cells. The International Society for Cellular Therapy position statement. Cytotherapy. 2006;8(4):315–7.
4. Barry FP, Murphy JM. Mesenchymal stem cells: clinical applications and biological characterization. Int J Biochem Cell Biol. 2004;36(4):568–84.
5. Pittenger MF, Mackay AM, Beck SC, Jaiswal RK, Douglas R, Mosca JD, et al. Multilineage potential of adult human mesenchymal stem cells. Science. 1999;284(5411):143–7.
6. Lai RC, Yeo RWY, Tan SS, Zhang B, Yin Y, Sze NSK, et al. Mesenchymal stem cell exosomes: the future msc-based therapy? In: Mesenchymal stem cell therapy. Totowa: Humana Press; 2013. p. 39–61.
7. Giordano A, Galderisi U, Marino IR. From the laboratory bench to the patient's bedside: an update on clinical trials with mesenchymal stem cells. J Cell Physiol. 2007;211(1):27–35.
8. Afanasyev BV, Elstner EE, Zander AR. A. J. Friedenstein, founder of the mesenchymal stem cell concept. Cell Ther Transplant. 2009;1(3):35–8.
9. Friedenstein AJ, Chailakhyan RK, Latsinik NV, Panasyuk AF, Keiliss-Borok IV. Stromal cells responsible for transferring the microenvironment of the hemopoietic tissues. Cloning in vitro and retransplantation in vivo. Transplantation. 1974;17(4):331–40.
10. Friedenstein AJ, Petrakova KV, Kurolesova AI, Frolova GP. Heterotopic of bone marrow. Analysis of precursor cells for osteogenic and hematopoietic tissues. Transplantation. 1968;6(2):230–47. [cited 2018 Aug 5].
11. Ullah I, Subbarao RB, Rho GJ. Human mesenchymal stem cells - current trends and future prospective. Biosci Rep. 2015;35(2):e00191.
12. Hass R, Kasper C, Böhm S, Jacobs R. Different populations and sources of human mesenchymal stem cells (MSC): a comparison of adult and neonatal tissue-derived MSC. Cell Commun Signal. 2011;9(1):12.
13. Brunstein CG, Setubal DC, Wagner JE. Expanding the role of umbilical cord blood transplantation. Br J Haematol. 2007;137(1):20–35.
14. Phinney DG, Prockop DJ. Concise review: mesenchymal stem/multipotent stromal cells: the state of transdifferentiation and modes of tissue repair-current views. Stem Cells. 2007;25(11):2896–902.

15. Vizoso F, Eiro N, Cid S, Schneider J, Perez-Fernandez R. Mesenchymal stem cell secretome: toward cell-free therapeutic strategies in regenerative medicine. Int J Mol Sci. 2017;18(9):1852.
16. Zimmerlin L, Park TS, Zambidis ET, Donnenberg VS, Donnenberg AD. Mesenchymal stem cell secretome and regenerative therapy after cancer. Biochimie. 2013;95(12):2235–45.
17. Caplan AI, Dennis JE. Mesenchymal stem cells as trophic mediators. J Cell Biochem. 2006;98(5):1076–84.
18. McNamara LE, Sjöström T, Meek RMD, Oreffo ROC, Su B, Dalby MJ, et al. Metabolomics: a valuable tool for stem cell monitoring in regenerative medicine. J R Soc Interface. 2012;9(73):1713–24.
19. Minai-Tehrani A, Jafarzadeh N, Gilany K. Metabolomics: a state-of-the-art technology for better understanding of male infertility. Andrologia. 2016;48(6):609–16.
20. Fiehn O. Metabolomics — the link between genotypes and phenotypes. In: Functional genomics. Dordrecht: Springer; 2002. p. 155–71.
21. Goodacre R. Metabolomics of a superorganism. J Nutr. 2007;137(1):259S–66S.
22. Griffin JL. The Cinderella story of metabolic profiling: does metabolomics get to go to the functional genomics ball? Philos Trans R Soc Lond Ser B Biol Sci. 2006;361(1465):147–61.
23. Wishart DS, Feunang YD, Marcu A, Guo AC, Liang K, Vázquez-Fresno R, et al. HMDB 4.0: the human metabolome database for 2018. Nucleic Acids Res. 2018;46(D1):D608–17.
24. Dando I, Dalla Pozza E, Biondani G, Cordani M, Palmieri M, Donadelli M. The metabolic landscape of cancer stem cells. IUBMB Life. 2015;67(9):687–93.
25. Shyh-Chang N, Ng H-H. The metabolic programming of stem cells. Genes Dev. 2017;31(4):336–46.
26. Murgia A, Mancuso L, Manis C, Caboni P, Cao G. GC-MS metabolomics analysis of mesenchymal stem cells treated with copper oxide nanoparticles. Toxicol Mech Methods. 2016;26(8):611–9.
27. Mastrangelo A, Panadero MI, Pérez LM, Gálvez BG, García A, Barbas C, et al. New insight on obesity and adipose-derived stem cells using comprehensive metabolomics. Biochem J. 2016;473(14):2187–203.
28. Ivanova G, Pereira T, Caseiro AR, Georgieva P, Maurício AC. Metabolomic and proteomic analysis of the mesenchymal stem cells' secretome. In: Metabolomics - fundamentals and applications. London: InTech; 2016.
29. West FD, Henderson WM, Yu P, Yang J-Y, Stice SL, Smith MA. Metabolomic response of human embryonic stem cell-derived germ-like cells after exposure to steroid hormones. Toxicol Sci. 2012;129(1):9–20.
30. Kim J-S, Kim E-J, Kim H-J, Yang J-Y, Hwang G-S, Kim C-W. Proteomic and metabolomic analysis of H2O2-induced premature senescent human mesenchymal stem cells. Exp Gerontol. 2011;46(6):500–10.
31. Cezar GG, Quam JA, Smith AM, Rosa GJM, Piekarczyk MS, Brown JF, et al. Identification of small molecules from human embryonic stem cells using metabolomics. Stem Cells Dev. 2007;16(6):869–82.
32. Wang G, Cao K, Liu K, Xue Y, Roberts AI, Li F, et al. Kynurenic acid, an IDO metabolite, controls TSG-6-mediated immunosuppression of human mesenchymal stem cells. Cell Death Differ. 2018;25(7):1209–23.
33. Pan Z, Raftery D. Comparing and combining NMR spectroscopy and mass spectrometry in metabolomics. Anal Bioanal Chem. 2007;387(2):525–7.
34. Cui L, Lu H, Lee YH. Challenges and emergent solutions for LC-MS/MS based untargeted metabolomics in diseases. Mass Spectrom Rev. 2018;37(6):772–92.
35. Brion C, Miller S. Regulated and constitutive secretion. Differential effects of protein synthesis arrest on transport of glycosaminoglycan chains to the two secretory pathways. J Biol Chem. 1992;267(3):1477–83.
36. Kilroy GE, Foster SJ, Wu X, Ruiz J, Sherwood S, Heifetz A, et al. Cytokine profile of human adipose-derived stem cells: expression of angiogenic, hematopoietic, and pro-inflammatory factors. J Cell Physiol. 2007;212(3):702–9.

37. Lee MJ, Kim J, Kim MY, Bae Y-S, Ryu SH, Lee TG, et al. Proteomic analysis of tumor necrosis factor-α-induced secretome of human adipose tissue-derived mesenchymal stem cells. J Proteome Res. 2010;9(4):1754–62.
38. Kupcova Skalnikova H. Proteomic techniques for characterisation of mesenchymal stem cell secretome. Biochimie. 2013;95(12):2196–211.
39. Camargo M, Intasqui P, Bruna de Lima C, Montani DA, Nichi M, Pilau EJ, et al. MALDI-TOF fingerprinting of seminal plasma lipids in the study of human male infertility. Lipids. 2014;49(9):943–56.
40. Bligh EG, Dyer WJ. A rapid method of total lipid extraction and purification. Can J Biochem Physiol. 1959;37(8):911–7.
41. Castiglione F, Ferro M, Mavroudakis E, Pellitteri R, Bossolasco P, Zaccheo D, et al. NMR metabolomics for stem cell type discrimination. Sci Rep. 2017;7(1):15808.
42. Gilany K, Moazeni Pouracil RS, Reza Sadeghi M. Fourier transform infrared spectroscopy: a potential technique for noninvasive detection of spermatogenesis. Avicenna J Med Biotechnol. 2014;6(1):47–52.
43. Gilany K, Jafarzadeh N, Mani-Varnosfaderani A, Minai-Tehrani A, Sadeghi MR, Darbandi M, et al. Metabolic fingerprinting of seminal plasma from non-obstructive azoospermia patients: positive versus negative sperm retrieval. J Reprod Infertil. 2018;19(2):109–14.
44. Jafarzadeh N, Mani-Varnosfaderani A, Minai-Tehrani A, Savadi-Shiraz E, Sadeghi MR, Gilany K. Metabolomics fingerprinting of seminal plasma from unexplained infertile men: a need for novel diagnostic biomarkers. Mol Reprod Dev. 2015;82:150.
45. Gilany K, Moazeni-Pourasil RS, Jafarzadeh N, Savadi-Shiraz E. Metabolomics fingerprinting of the human seminal plasma of asthenozoospermic patients. Mol Reprod Dev. 2014;81(1):84–6.
46. Zhang J, Rector J, Lin JQ, Young JH, Sans M, Katta N, et al. Nondestructive tissue analysis for ex vivo and in vivo cancer diagnosis using a handheld mass spectrometry system. Sci Transl Med. 2017;9(406):eaan3968.
47. Johnson CH, Ivanisevic J, Benton HP, Siuzdak G. Bioinformatics: the next frontier of metabolomics. Anal Chem. 2015;87(1):147–56.
48. Guijas C, Montenegro-Burke JR, Domingo-Almenara X, Palermo A, Warth B, Hermann G, et al. METLIN: a technology platform for identifying knowns and unknowns. Anal Chem. 2018;90(5):3156–64.

Different Gene Expression Profile of Mesenchymal Stem Cells from Various Sources

Babak Arjmand, Negar Ranjbaran, Fatemeh Khatami, and Mehrdad Hashemi

Abstract Mesenchymal stem cells can be derived from different tissues of the body. It is important to select the suitable tissue that has the main characteristics of these types of cells. As we know, behind each specific property, a specific gene is expressed and the molecular process is in progress. Evaluating morphology, surface markers, potency, gene expression, and differentiation potential can be helpful to choose the most appropriate cellular source. These data considered a higher expression of chondrogenesis markers in bone marrow-derived and osteogenesis markers in umbilical cord blood-derived mesenchymal stem cells, and also an increase in angiogenesis and proliferation potential of umbilical cord mesenchymal stem cells. Comprehensive information of gene expression profile of mesenchymal stem cells can make it possible to use the mesenchymal stem cells in cell therapy, pharmacological tests, tissue engineering, and also has a positive role in omics studies.

Keywords Bone marrow · Gene expression · Mesenchymal stem cells · Placenta · Transcriptome profiles

B. Arjmand (✉)
Cell Therapy and Regenerative Medicine Research Center, Endocrinology and Metabolism Molecular-Cellular Sciences Institute, Tehran University of Medical Sciences, Tehran, Iran

Metabolomics and Genomics Research Center, Endocrinology and Metabolism Molecular-Cellular Sciences Institute, Tehran University of Medical Sciences, Tehran, Iran
e-mail: barjmand@sina.tums.ac.ir

N. Ranjbaran · M. Hashemi
Department of Genetics, Tehran Medical Sciences, Islamic Azad University, Tehran, Iran

F. Khatami
Chronic Disease Research Center, Endocrinology and Metabolism Population Sciences Institute, Tehran University of Medical Sciences, Tehran, Iran

B. Arjmand (ed.), *Genomics, Proteomics, and Metabolomics*, Stem Cell Biology and Regenerative Medicine, https://doi.org/10.1007/978-3-030-27727-7_4

1 Stem Cells

Undifferentiated cells with special features such as the potential of self-renewal and differentiating into multi-lineages are called "stem cells" which are classified based on their potency and origins.

1.1 Classification of Stem Cells

1.1.1 Potency-Based Classification of Stem Cells

Classification of stem cells is based on their ability to differentiate into several cell types, including totipotent, pluripotent, multipotent, oligopotent, and unipotent [1].

1. Totipotent: All types of cells that form the body are the result of the differentiation of these types of stem cells such as zygote and morula [1, 2].
2. Pluripotent: Stem cells can be differentiated into almost all cell types, e.g., differentiation of embryonic stem cells into three cell lineages (mesoderm, endoderm, and ectoderm) [1–3] (www.differencebetween.net/science/difference-between-totipotent-and-pluripotent; accessed 9/21/2018).
3. Multipotent: Familiar cell types are outcomes of these differentiated stem cells, such as blood stem cells and adult stem cells that can differentiate into monocyte and lymphocyte [1] (www.en.wikipedia.org/wiki/Cell_potency#Totipotency; accessed 9/21/2018).
4. Oligopotent: Some specific cell lineages concluded by these stem cells' differentiation potential, e.g., lymphoid and myeloid stem cells [1, 4].
5. Unipotent: Differentiation of stem cells with self-renewal potency into one specific cell type, e.g., muscle stem cells [1, 4].

1.1.2 Origin-Based Classification of Stem Cells

Classification of stem cells based on their provenance is defined as the embryonic stem cells (ESCs), induced pluripotent stem cells (iPSCs), and adult stem cells [1, 5].

1.1.2.1 Embryonic Stem Cells

Pluripotent stem cells isolating from the inner cell mass of blastocyst are called "embryonic stem cells" [1, 2, 6–8]. Abilities of pluripotency, indestructibility [1], and adaptability are hallmarks of these types of stem cells [2]. However, researchers encounter with some barriers when using ESCs in therapy include rejection of cell engraftment because of incoherence between the donor cells and the recipients, the ethical concerns about embryo death during sampling [6], and presentation of teratomas after transplantation [9]. The ESCs resemble another class of stem cells due to their pluripotency and gene expression [10], the "induced pluripotent stem cells."

1.1.2.2 Induced Pluripotent Stem Cells

Reprogrammed differentiated somatic cells that the expression of their genes was converted into pluripotent cells are called "induced pluripotent stem cells" [1, 6, 11]. The potential to differentiate into all three cell lineages such as ectoderm, mesoderm, and endoderm *in vitro*, and the potential for infinite self-renewal are important in cell therapy. The specific set of genes (Sox2,[1] Oct4,[2] and Nanog[3]) and signaling molecules play a role in pluripotency of these cells [6].

1.1.2.3 Adult Stem Cells

Multipotent or totipotent stem cells isolated from undifferentiated cells multiply by cell division to replenish dying cells and regenerate damaged tissues. Body tissues such as umbilical cord are the sources of adult stem cells [1, 2]. In this classification, mesenchymal stem cells are categorized as the subtype of adult stem cells (Fig. 1) [5].

2 Mesenchymal Stem Cells

In the late 1990s for the first time, the mesenchymal stem cells were isolated from bone marrow of rats and pigs and described as marrow stromal cells [12, 13]. Investigations revealed the presence of mesenchymal stem cells in different body tissues [14]. MSCs[4] were isolated from various tissue sources of body such as bone marrow (BM), adipose tissue (AT), placenta (PL), Wharton's jelly (WJ), umbilical cord blood (UCB), peripheral blood (PB), amniotic fluid, skeletal muscle, tendons, skin, lung, liver, synovium, fetal blood, and dental pulp [13, 15–18].

---

[1] SRY-box 2.

[2] Octamer-binding transcription factor 4.

[3] Nanog homeobox.

[4] Mesenchymal stem cells.

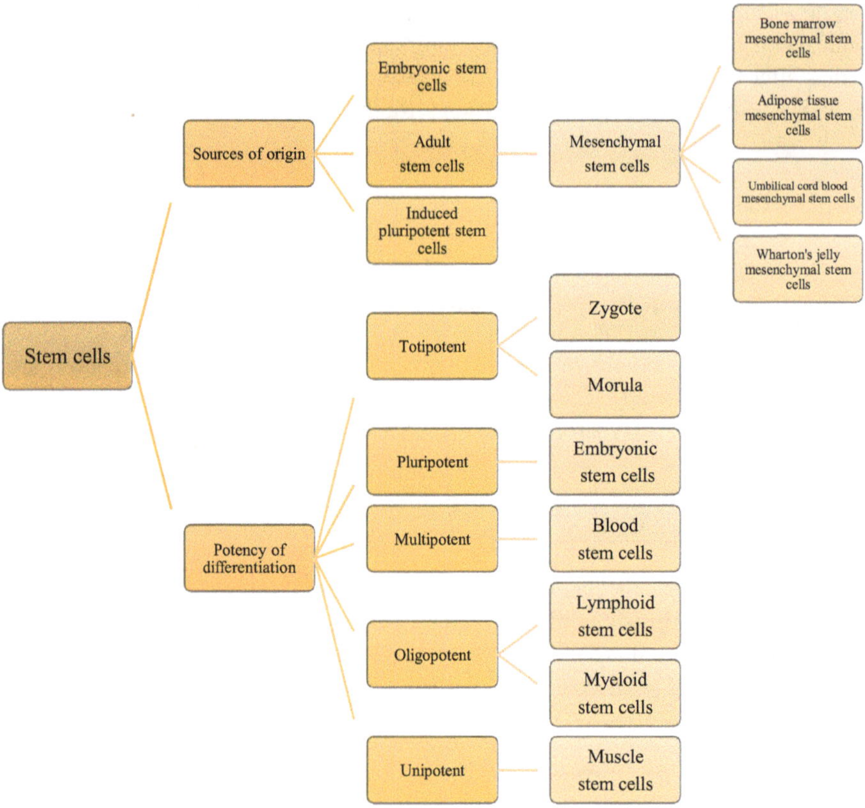

Fig. 1 Classification of stem cells based on their provenance and potency. Embryonic stem cells, adult stem cells, and induced pluripotent stem cells are the three types of stem cell's origins. Classification based on their differentiation potential contains clusters of totipotent, pluripotent, multipotent, oligopotent, and unipotent stem cells

2.1 Sources of Mesenchymal Stem Cells

2.1.1 Bone Marrow

The central cavity of bones composed of gelatinous tissue with blood vessels. Bone marrow consists of myeloid tissue (red marrow) and fatty tissue (yellow marrow). MSCs derived from adipose tissue generate bone and fat. Hematopoietic stem cells derived from the myeloid tissue of bone marrow form the blood stem cells [19] (www.medicalnewstoday.com/articles/285666.php; accessed 10/28/2018). Friedenstein et al. in 1976 [20] investigated that these multipotent stem cells can be differentiated into osteocytes, adipocytes, and chondrocytes in vitro. Adherent to plastic and expressing cell surface markers (CD44, CD105, CD106, CD73, CD29, CD166, CD90, CD117, STRO-1, Sca-1) are their other characteristics [2, 17, 18, 21]. Some genes affect differentiation and cell growth of bone marrow mesenchymal stem cells by their expression such as the calcium-dependent phospholipid-binding protein "annexin A2" [22–24]. Describing BM-MSCs'[5] features is important. Painful harvest, reverse effect of age on efficiency, and deficiency in MSCs derived from bone marrow are the limitations of using this source for clinical applications. To overcome these limitations, adipose tissue have been strongly suggested as a valuable source of MSCs [18, 25, 26].

2.1.2 Adipose Tissue

The loos connective tissue consists of adipocytes, stem cells, endothelial cells, immune cells, and neuronal cells [27–29]. Adipose tissue is found in soles, palms, around joints, heart, muscles [27], so the greater amount of it can be found in the body [30]. Based on histological and molecular structure, adipose tissue consists of two parts: white adipose tissue and brown adipose tissue [27, 31]. WAT[6] stores energy and systematizes metabolism. BAT[7] is more functional in newborns and produces body heat. MSCs mostly isolated from WAT because of the greater amount, self renewal capacity and potency its MSCs [27]. Ad-MSCs[8] can be isolated from the waste product of liposuction, the lipoaspirate. The International Fat Applied Technology Society has settled distinct "adipose-derived stem cells" instead of all terms that include pre-adipocytes, multipotent adipose-derived stem cells, adipose-derived adult stem cells, stromal cells, and lipoaspirate cells. Differentiation potential into adipocytes, osteocytes, myocytes, and chondrocytes, easy to culture, rapid growth and impress on angiogenesis, influence of transplantation of fats, and regeneration of damaged tissues are significance characteristics of Ad-MSCs [25].

2.1.3 Umbilical Cord Blood

Umbilical cord consists of a vein and two arteries surrounded by Wharton's jelly, which connects placenta to fetus, providing fetal nutrition and eliminating wastes [32] (www.medical-dictionary.thefreedictionary.com/umbilical+cord; accessed 11/14/2018).

Umbilical cord blood (UCB) is transmitted by veins and contains circulating stem cells that can be isolated and characterized as a source of stem cells for research and various clinical applications [25, 32–34]. UCB can be isolated without using surgical invasive procedure [33]. Umbilical cord-driven mesenchymal stem cells are multipotent stem cells with the potential of differentiating into neurocytes, cardiomyocytes, and hepatocytes [25, 32]. Low expression of the beta-2-microglobulin in UCB along with the high production of interleukin-10 reduces the rejection of the

[5] Bone marrow-derived mesenchymal stem cells.

[6] White adipose tissue.

[7] Brown adipose tissue.

[8] Adipose tissue-derived mesenchymal stem cells.

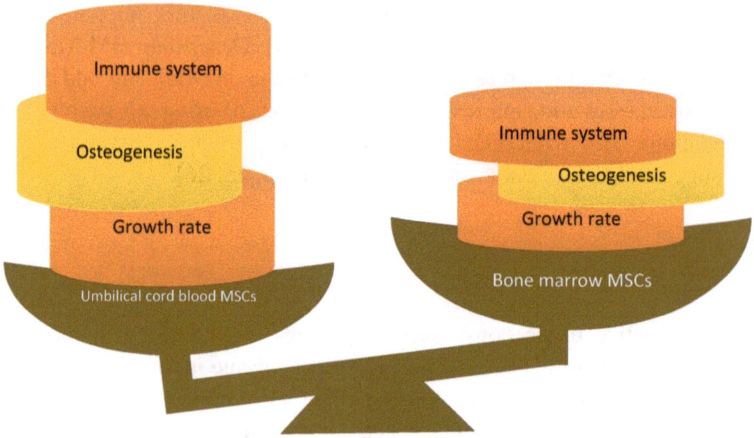

Fig. 2 Comparison between different sources such as bone marrow mesenchymal stem cells and umbilical cord blood mesenchymal stem cells shows higher gene markers expression of immune system, osteogenesis, and growth rate in UCB-MSCs than BM-MSCs

graft, and they can be stored to use for a longer time [2]. However, isolation of UCB from placenta has difficulties, and it is important to ensure its efficacy after a long period of storing [25]. The comparison between UCB-MSCs and BM-MSCs shows more expression of genes that participate in osteogenesis, immune modulation, and multilineage differentiation in UCB-MSCs than BM-MSCs [35]. BM-MSCs have lower proliferation than UCB-MSCs. Also, analyzing the endothelial specific markers shows greater angiogenesis potential in UC-MSCs than BM-MSCs [36]. The growth rate of the UCB-MSCs[9] is greater than that of Ad-MSCs and BM-MSCs [15] (Fig. 2). There are some shared expressed genes between BM-MSCs and UCB-MSCs such as collagens, annexin A2, osteonectin, and galectin 1 [35, 37].

2.1.4 Wharton's Jelly

Wharton's jelly is the gelatinous connective tissue that preserves umbilical cord vein and arteries. Due to the presence of collagen fibrils and glycoprotein microfibrils, Wharton's Jelly acts as a shield against pressure. These matrix cells of umbilical cord are the rich source of MSCs. They are waste product of parturition that can be isolated easily and painlessly [32, 38–40]. The MSCs that are isolated painlessly from Wharton's jelly have the ability to differentiate into neurocytes, adipocytes, osteocytes, chondrocytes, cardiomyocytes in vitro [30, 41] and express cell markers such as c-kit that can proliferate for a longer time [2]. High growth rates, multipotency ability and no impact on teratomas consider the Wharton's jelly-derived MSCs as an option for using in clinical applications [26].

[9] Umbilical cord blood mesenchymal stem cells

Table 1 Sources and mechanisms of MSCs based on gene expressions

References	Sources	Coding gene	Gene location in Homo sapiens (Human)	Effects
[22]	Bone marrow mesenchymal stem cells	ANXA2 (Annexin A2)	15q22.2	Regulate osteogenesis and proliferation
[45]	Adipose-derived mesenchymal stem cells	LPL (lipoprotein lipase)	8p21.3	Upregulate lipid metabolism in adipogenesis
[2]	Umbilical cord blood mesenchymal stem cells	IL-10 (interleukin 10)	1q32.1	Reduce rejection of grafts
[42]	Wharton's jelly mesenchymal stem cells	NTN1 (Netrin-1)	17p13.1	Regulate angiogenesis
[43]		S100A6 (S100 calcium binding protein A6)	1q21.3	Enhance cell adhesion Inhibit proliferation
[41]		IGF-1 (insulin-like growth factor 1)	12q23.2	Enhance chondrogenesis

Various genes are expressed in Wharton's jelly, which changes specific process in the MSCs including NTN-1,[10] S100A6, IGF1, IGF2. Netrin-1, the ligand of laminin-like proteins, takes action with DCC/Neogenin-1 and UNC5 that regulate angiogenesis [42]. WJ-MSCs[11] release S100A6 in extracellular matrix. This calcium-binding protein enhances adhesion of cells, therefore stops proliferation (Table 1) [43, 44].

Insulin growth factor 1 functions as an enhancer in the gene expression. IGF1[12] in WJ-MSCs increases chondrocyte differentiation by affecting the expression of chondrocyte markers including COL2 and SOX9. IGF2[13] is used during the differentiation stage for regulating glucose and non-differentiation stage for motivating growth [32, 41].

2.2 *Characteristics of Mesenchymal Stem Cells*

Some special features that can be used in distinguishing MSCs from other cells include adherent to plastic which makes MSCs to easily isolate, expand, and grow in vitro in culture dishes [13, 16–18, 46], expression of cell surface markers [12, 13,

[10] Netrin-1.

[11] Wharton's jelly-derived mesenchymal stem cells.

[12] Insulin growth factor 1.

[13] Insulin growth factor 2.

16, 17], potential to differentiate into cell lineages [12, 13, 15, 17], self-renewal potency [12, 15, 17], multilineage potential [12, 13, 15–17], proliferation [12, 13, 15, 17, 46], and immunomodulatory properties [15]. The spindle-shaped MSCs simulate fibroblast cells [13, 18, 37]. Senescence increases debris and granulates the cytoplasm [47]. Cellular senescence occurs in the UCB-MSCs later than other MSCs, which is followed by a lower expression of senescence-related proteins, while Ad-MSCs and BM-MSCs display greater expression of senescence-related proteins [33]. Researches show a reduction in the number of MSCs in elderlies [18]. Although working on cell therapy of ischemic cardiomyopathy debates that there is no difference in age of patients [48].

2.2.1 Cell Surface Markers

Expression of cell surface markers such as adhesion molecules and matrix receptors (CD105, CD44), integrin markers (CD29, CD51), CD73, CD90 and lack of expression of hematopoietic lineage markers (CD45, CD34), CD14 (CD11b), CD19, and HLA-DR have been described by the International Society for Cell Therapy (ISCT, http://www.celltherapysociety.org/) as identification method [13, 16, 17, 32, 49]. Also expression of mesenchymal stem cell markers (SH2, SH3) can be discussed. Cell surface markers are evaluated by flow cytometry [32].

Comparing immunophenotyping of BM-MSCs, UCB-MSCs, and Ad-MSCs shows that the expression of CD44, CD90, CD105, CD45, CD73, and CD29 is higher in BM-MSCs than in other sources. The expression of CD34 in UCB-MSCs and CD14 in Ad-MSCs was significant [15].

2.2.2 Differentiation

Appropriate culture conditions (cell density, cytokines, and growth factors) [18] motivate mesenchymal stem cells to differentiate into mesodermal lineage such as adipocytes, osteocytes, chondrocytes, tenocytes, and cardiomyocytes [5, 12, 13, 16, 17], ectodermal lineages that include neurocytes, and endodermal lineages as hepatocytes [5, 13, 16].

DKK1 use in intervene of protein interactions increases its effect on proliferation and cell growth (https://www.ncbi.nlm.nih.gov/gene/22943; accessed 10/28/2018). Secretion of Dkk1 in low density of MSCs shows undifferentiated cells. When cells collide, DKK1 affects rescind by Wnt-5a expression.

Researchers also discuss the relation of the differentiation potential with the source of MSCs. MSCs treated by dexamethasone, indomethacin, and isobutyl methyl xanthine (IBMX) differentiate into adipocytes. Certainty of differentiation confirmed by various ways includes Oil red O staining that shows the red lipid droplets and expression of specific markers, such as peroxisome proliferation-activated receptor $\gamma 2$ (PPAR$\gamma 2$), lipoprotein lipase (LPL), and the fatty acid binding protein 4 (FABP4) [18, 50, 51]. BM-MSCs have lower adipogenesis potential than osteogenesis [45, 52]. Genes that participate in different metabolic sections of cholesterol and lipid metabolism, such as FABP4, PPARγ, LPL, APOD, and LEP, are more expressed in Ad-MSCs, leading to greater adipogenesis potential than BM-MSCs [45]. MSCs treated by 5-azacytidine, Troponin I, and F-actin differentiate into cardiomyocytes. A distinction can be made by examining the expression of the cardiomyocyte markers such as *N*-cadherin and cardiac troponin I, connexin43 [32]. Treating MSCs by medium that contains dexamethasone, ascorbic acid, and β-glycerophosphate induced osteogenic differentiation [32, 53]. Appearance of RUNX2, Collagen I, was followed by osteogenesis [53]. During the process of differentiation, the antigen expression is convertible, such as CD166 antigen (activated leukocyte cell adhesion molecule) that vanishes during osteogenesis [18]. Alizarin red staining used to show granules contains calcium [51]. Measuring osteoblastic markers and staining result that BM-MSCs and WJ-MSCs differentiate into osteocyte less than Ad-MSCs and UCB-MSCs (Fig. 3). Examining the osteogenesis in these sources concluded that Ad-MSCs, UCB-MSCs, and WJ-MSCs are better superseded instead of BM-MSCs [30, 54]. Chondrogenesis is the result of expression of collagen X, and chondroadherin genes [55], collagen II [41, 55], SOX9 [41, 51] treating MSCs with differentiation medium that contains sodium pyruvate, ascorbic acid-2-phosphate, dexamethasone, and TGF-β1 [51, 55, 56]. Alcian blue

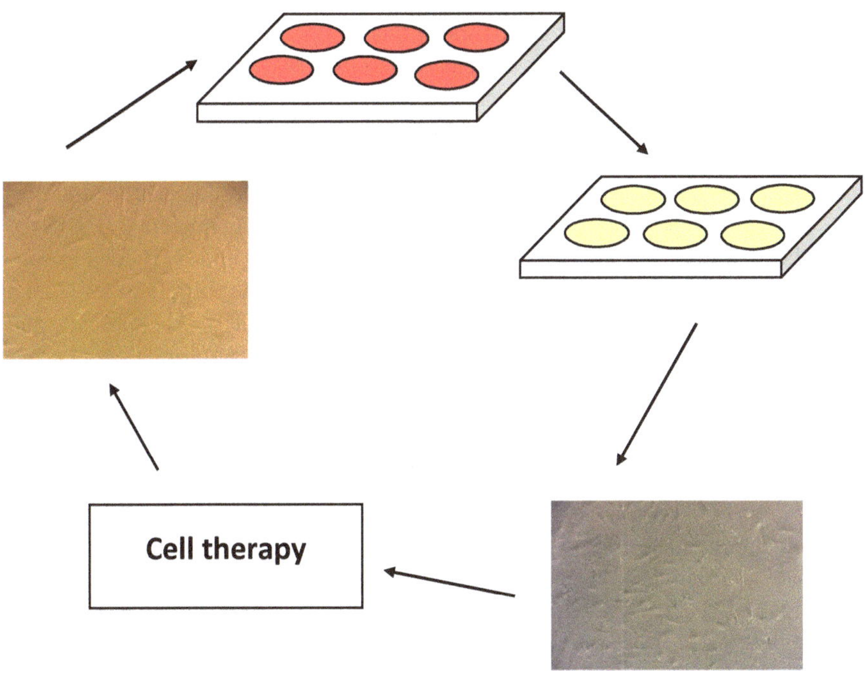

Fig. 3 Differentiation process of adipose-derived mesenchymal stem cells into osteocytes. Ad-MSCs have high osteogenesis potential

and safranin O staining were used for the detection of chondrogenesis [33, 56]. BM-MSCs have higher chondrocyte differentiation ability than Ad-MSCs, due to inspiration of IGF1 and TGFβ2 in COL2, COL1, and SOX9 [41].

2.2.3 Self-Renewal

Self-renewal is a capacity of cell division in MSCs. Cell division types include symmetric that creates two similar daughter stem cells with identical properties and asymmetric that generates differentiated MSCs with special effects [12].

2.2.4 Proliferation

Specific culture dishes provide conditions for MSCs proliferation to generate more stem cells [13]. While the cells are multiplying for more than 50 times in population, they have the ability to differentiate. Growth factors and cytokines can affect proliferation potency [18]. Expansion of MSCs *in vitro* is limited by the number of passages that cause deterioration by enhancing passages [12, 46]. The expression of chemokine receptor changes by multiplicity of passages, for example, BM-MSCs expressed CCR1, CCR9, CXCR4, and CXCR5 at early passage (P2). These molecules were absent at later passages (P12–16th). The cells could not transfer toward the chemokine attractants that corroborated this fact. Furthermore, reduction in the expression of adhesion molecules such as CD157, ICAM-1, ICAM-2, and VCAM-1 occurs with disappearing receptor expression. Some antigens can be expressed in early isolated MSCs while, they are absent in in vitro culture based on the culture conditions and passage number [18]. Comparison of MSCs from various sources shows a higher potency and shorter expantion time in Ad-MSCs than other counterparts [33]. That is because of the great presence of inhibin subunit beta B (INHBB), Ras-related dexamethasone induced 1 (RASD1), Zinc finger and BTB domain containing 16 (ZBTB16) genes in Ad-MSCs that participate in proliferation potential and cell growth and even in cell cycle [45]. Continuously, UCB-MSCs and WJ-MSCs have the highest and BM-MSCs have the lowest proliferation capacity [30].

2.2.5 Immunomodulatory Potential

The immunomodulatory potential of MSCs may be affected by condition of isolation and culturing [57]. Secretion of soluble factors for instance transforming growth factor-β, nitric oxide, interleukin1β, indole amine 2,3-dioxygenase, histocompatibility leucocyte antigen-G, prostaglandin E2, and contact between cells may affect the action of mediating immunomodulatory potential of MSCs [58].

3 Conclusion

Analyzing gene expression profiles shows the comparison between different sources of MSCs. Studies on protein-coding gene expression and noncoding RNAs manifest the obscurity about ontogeny and biology of MSCs [35].

According to the origin of MSCs, expression of specific genes may be greater than others. The genes correlated with osteogenesis and adipogenesis be expressed in BM-MSCs, and the genes correlated with angiogenesis, neurogenesis, and cell adhesion were expressed in UCB-MSCs [35, 37]. The gene expression profiles of transforming growth factor-α induced, transgelin, galectin 1, laminin receptor 1, and profilin 1 demonstrated the similarities between different origin of MSCs [37].

Due to recent advances in therapy, MSCs have attracted attention to the characteristics in therapeutic application and regenerative medicine [15, 17, 35, 37, 46]. Analyzing gene expression of MSCs helps to choose proper cells for cell therapy. Depending on the genes that expressed more, MSCs are used for the treatment of particular part of the body. UC-MSCs are widely used in treatments which need revascularization than BM-MSCs [37]. Recently, MSCs are used in the treatment of cancer, chronic wounds, and cell therapy. Research on their unknown features is still ongoing [15, 17, 35, 37, 46]. Except for BM-MSCs, other MSCs are isolated from other tissues by non-invasive methods [30]. The barriers that make us to select the most suitable source for treatment include ethical concerns, painful and invasive methods of isolating, and low amount of MSC's pool. To overcome these limitations, researchers investigated the expression of MSCs to carry out molecular analysis [18, 46]. Different methods including Reverse Transcription Polymerase Chain Reaction (RT-PCR), Real-time Quantitative Polymerase Chain Reaction (RT-Q-PCR), Serial Analysis of Gene Expression (SAGE), and Microarray were used for molecular analysis [35, 37, 59]. Furthermore, analysis of genes such as collagen1A, MMP2, TAGLN2, VIM, SPARC, interleukin (IL)-8, LGALS1 was carried out [35, 37].

References

1. Kalra K, Tomar P. Stem cell: basics, classification and applications. Am J Phytomed Clin Ther. 2014;2(7):919–30.
2. Bongso A, Lee EH. Stem cells: their definition, classification and sources. In: Stem cells: from bench to bedside. Singapore: World Scientific; 2005. p. 1–13.
3. Manganelli G, Fico A, Filosa S. Embryonic stem cells: from blastocyst to in vitro differentiation. In: Methodological advances in the culture, manipulation and utilization of embryonic stem cells for basic and practical applications. London: InTech; 2011.
4. Kolios G, Moodley Y. Introduction to stem cells and regenerative medicine. Respiration. 2013;85(1):3–10.
5. Ullah I, Subbarao RB, Rho GJ. Human mesenchymal stem cells - current trends and future prospective. Biosci Rep. 2015;35(2):e00191.

6. Medvedev S, Shevchenko A, Zakian S. Induced pluripotent stem cells: problems and advantages when applying them in regenerative medicine. Acta Naturae. 2010;2(2):18–28.
7. Cho S-G, Honguntikar S, Lee HJ. Application of magnet-based nanofection in embryonic stem cell research. In: Methodological advances in the culture, manipulation and utilization of embryonic stem cells for basic and practical applications. London: InTech; 2011.
8. Keller KC, zur Nieden NI. Osteogenesis from pluripotent stem cells: neural crest or mesodermal origin? In: Embryonic stem cells-differentiation and pluripotent alternatives. Singapore: InTech; 2011.
9. Arjmand B, Goodarzi P, Falahzadeh K, Aghayan HR, Rahim F, Mohamadi-Jahani F, et al. GMP-compliant perinatal tissue-derived stem cells. In: Perinatal tissue-derived stem cells. Berlin: Springer; 2016. p. 189–213.
10. Yan X, Xu N, Meng C, Wang B, Yuan J, Wang C, et al. Generation of induced pluripotent stem cells from human mesenchymal stem cells of parotid gland origin. Am J Transl Res. 2016;8(2):419.
11. Ko SH, Nauta A, Wong V, Glotzbach J, Gurtner GC, Longaker MT. The role of stem cells in cutaneous wound healing: what do we really know? Plast Reconstr Surg. 2011;127:10S–20S.
12. Elahi KC, Klein G, Avci-Adali M, Sievert KD, MacNeil S, Aicher WK. Human mesenchymal stromal cells from different sources diverge in their expression of cell surface proteins and display distinct differentiation patterns. Stem Cells Int. 2016;2016:5646384.
13. Kobolak J, Dinnyes A, Memic A, Khademhosseini A, Mobasheri A. Mesenchymal stem cells: identification, phenotypic characterization, biological properties and potential for regenerative medicine through biomaterial micro-engineering of their niche. Methods. 2016;99:62–8.
14. Nishikiori R, Watanabe K, Kato K. Antibody arrays for quality control of mesenchymal stem cells. ACS Appl Mater Interfaces. 2015;7(30):16828–36.
15. Heo JS, Choi Y, Kim HS, Kim HO. Comparison of molecular profiles of human mesenchymal stem cells derived from bone marrow, umbilical cord blood, placenta and adipose tissue. Int J Mol Med. 2016;37(1):115–25.
16. Wagner W, Wein F, Seckinger A, Frankhauser M, Wirkner U, Krause U, et al. Comparative characteristics of mesenchymal stem cells from human bone marrow, adipose tissue, and umbilical cord blood. Exp Hematol. 2005;33(11):1402–16.
17. Roson-Burgo B, Sanchez-Guijo F, Del Canizo C, De Las Rivas J. Transcriptomic portrait of human mesenchymal stromal/stem cells isolated from bone marrow and placenta. BMC Genomics. 2014;15:910.
18. Bobis S, Jarocha D, Majka M. Mesenchymal stem cells: characteristics and clinical applications. Folia Histochem Cytobiol. 2006;44(4):215–30.
19. Wong VW, Rustad KC, Longaker MT, Gurtner GC. Tissue engineering in plastic surgery: a review. Plast Reconstr Surg. 2010;126(3):858–68.
20. Friedenstein A, Latzinik N, Grosheva A, Gorskaya U. Marrow microenvironment transfer by heterotopic transplantation of freshly isolated and cultured cells in porous sponges. Exp Hematol. 1982;10(2):217–27.
21. Kern S, Eichler H, Stoeve J, Klüter H, Bieback K. Comparative analysis of mesenchymal stem cells from bone marrow, umbilical cord blood, or adipose tissue. Stem cells. 2006;24(5):1294–301.
22. Ye N-S, Zhang R-L, Zhao Y-F, Feng X, Wang Y-M, Luo G-A. Effect of 5-azacytidine on the protein expression of porcine bone marrow mesenchymal stem cells in vitro. Genomics Proteomics Bioinformatics. 2006;4(1):18–25.
23. Gilmore WS, Olwill S, McGlynn H, Alexander HD. Annexin A2 expression during cellular differentiation in myeloid cell lines. London: Portland Press Limited; 2004.
24. Jacovina AT, Deora AB, Ling Q, Broekman MJ, Almeida D, Greenberg CB, et al. Homocysteine inhibits neoangiogenesis in mice through blockade of annexin A2–dependent fibrinolysis. J Clin Invest. 2009;119(11):3384–94.
25. Locke M, Windsor J, Dunbar PR. Human adipose-derived stem cells: isolation, characterization and applications in surgery. ANZ J Surg. 2009;79(4):235–44.

26. Venugopal P, Balasubramanian S, Majumdar AS, Ta M. Isolation, characterization, and gene expression analysis of Wharton's jelly-derived mesenchymal stem cells under xeno-free culture conditions. Stem Cells Cloning. 2011;4:39–50.
27. Berry DC, Stenesen D, Zeve D, Graff JM. The developmental origins of adipose tissue. Development. 2013;140(19):3939–49.
28. Martinez-Santibañez G, Cho KW, Lumeng CN. Imaging white adipose tissue with confocal microscopy. Methods Enzymol. 2014;537:17–30.
29. Kershaw EE, Flier JS. Adipose tissue as an endocrine organ. J Clin Endocrinol Metabol. 2004;89(6):2548–56.
30. Kang B-J, Ryu H-H, Park SS, Koyama Y, Kikuchi M, Woo H-M, et al. Comparing the osteogenic potential of canine mesenchymal stem cells derived from adipose tissues, bone marrow, umbilical cord blood, and Wharton's jelly for treating bone defects. J Vet Sci. 2012;13(3):299–310.
31. Rashnonejad A, Ercan G, Gunduz C, Akdemir A, Tiftikcioglu YO. Comparative analysis of human UCB and adipose tissue derived mesenchymal stem cells for their differentiation potential into brown and white adipocytes. Mol Biol Rep. 2018;45(3):233–44.
32. Wang HS, Hung SC, Peng ST, Huang CC, Wei HM, Guo YJ, et al. Mesenchymal stem cells in the Wharton's jelly of the human umbilical cord. Stem Cells. 2004;22(7):1330–7.
33. Jin HJ, Bae YK, Kim M, Kwon S-J, Jeon HB, Choi SJ, et al. Comparative analysis of human mesenchymal stem cells from bone marrow, adipose tissue, and umbilical cord blood as sources of cell therapy. Int J Mol Sci. 2013;14(9):17986–8001.
34. Lee H-J, Jung J, Cho KJ, Lee CK, Hwang S-G, Kim GJ. Comparison of in vitro hepatogenic differentiation potential between various placenta-derived stem cells and other adult stem cells as an alternative source of functional hepatocytes. Differentiation. 2012;84(3):223–31.
35. Secco M, Moreira YB, Zucconi E, Vieira NM, Jazedje T, Muotri AR, et al. Gene expression profile of mesenchymal stem cells from paired umbilical cord units: cord is different from blood. Stem Cell Rev. 2009;5(4):387–401.
36. Chen M-Y, Lie P-C, Li Z-L, Wei X. Endothelial differentiation of Wharton's jelly–derived mesenchymal stem cells in comparison with bone marrow–derived mesenchymal stem cells. Exp Hematol. 2009;37(5):629–40.
37. Panepucci RA, Siufi JL, Silva WA Jr, Proto-Siquiera R, Neder L, Orellana M, et al. Comparison of gene expression of umbilical cord vein and bone marrow-derived mesenchymal stem cells. Stem Cells. 2004;22(7):1263–78.
38. Zhou C, Yang B, Tian Y, Jiao H, Zheng W, Wang J, et al. Immunomodulatory effect of human umbilical cord Wharton's jelly-derived mesenchymal stem cells on lymphocytes. Cell Immunol. 2011;272(1):33–8.
39. Lim JJ, Koob TJ. Placental cells and tissues: the transformative rise in advanced wound care. In: Worldwide wound healing-innovation in natural and conventional methods. London: InTech; 2016.
40. Mitchell KE, Weiss ML, Mitchell BM, Martin P, Davis D, Morales L, et al. Matrix cells from Wharton's jelly form neurons and glia. Stem Cells. 2003;21(1):50–60.
41. Widowati W, Afifah E, Mozef T, Sandra F, Rizal R, Amalia A, et al. Effects of insulin-like growth factor-induced Wharton jelly mesenchymal stem cells toward chondrogenesis in an osteoarthritis model. Iran J Basic Med Sci. 2018;21(7):745–52.
42. Prieto CP, Ortiz MC, Villanueva A, Villarroel C, Edwards SS, Elliott M, et al. Netrin-1 acts as a non-canonical angiogenic factor produced by human Wharton's jelly mesenchymal stem cells (WJ-MSC). Stem Cell Res Ther. 2017;8(1):43.
43. Jurewicz E, Goral A, Filipek A. S100A6 is secreted from Wharton's jelly mesenchymal stem cells and interacts with integrin beta1. Int J Biochem Cell Biol. 2014;55:298–303.
44. Jurewicz E, Kasacka I, Bankowski E, Filipek A. S100A6 and its extracellular targets in Wharton's jelly of healthy and preeclamptic patients. Placenta. 2014;35(6):386–91.
45. Liu TM, Martina M, Hutmacher DW, Hui JHP, Lee EH, Lim B. Identification of common pathways mediating differentiation of bone marrow-and adipose tissue-derived human mesenchymal stem cells into three mesenchymal lineages. Stem Cells. 2007;25(3):750–60.

46. Medeiros Tavares Marques JC, Cornelio DA, Nogueira Silbiger V, Ducati Luchessi A, de Souza S, Batistuzzo de Medeiros SR. Identification of new genes associated to senescent and tumorigenic phenotypes in mesenchymal stem cells. Sci Rep. 2017;7(1):17837.
47. Bonab MM, Alimoghaddam K, Talebian F, Ghaffari SH, Ghavamzadeh A, Nikbin B. Aging of mesenchymal stem cell in vitro. BMC Cell Biol. 2006;7(1):14.
48. Golpanian S, El-Khorazaty J, Mendizabal A, DiFede DL, Suncion VY, Karantalis V, et al. Effect of aging on human mesenchymal stem cell therapy in ischemic cardiomyopathy patients. J Am Coll Cardiol. 2015;65(2):125–32.
49. Kyurkchiev S, Shterev A, Dimitrov R. Assessment of presence and characteristics of multipotent stromal cells in human endometrium and decidua. Reprod Biomed Online. 2010;20(3):305–13.
50. Lee D-K, Yi T, Park K-E, Lee H-J, Cho Y-K, Lee SJ, et al. Non-invasive characterization of the adipogenic differentiation of human bone marrow-derived mesenchymal stromal cells by HS-SPME/GC-MS. Sci Rep. 2014;4:6550.
51. Ullah I, Subbarao Raghavendra B, Rho Gyu J. Human mesenchymal stem cells - current trends and future prospective. Biosci Rep. 2015;35(2):e00191.
52. Peng L, Jia Z, Yin X, Zhang X, Liu Y, Chen P, et al. Comparative analysis of mesenchymal stem cells from bone marrow, cartilage, and adipose tissue. Stem Cells Dev. 2008;17(4):761–74.
53. Langenbach F, Handschel J. Effects of dexamethasone, ascorbic acid and β-glycerophosphate on the osteogenic differentiation of stem cells in vitro. Stem Cell Res Ther. 2013;4(5):117.
54. Baksh D, Yao R, Tuan RS. Comparison of proliferative and multilineage differentiation potential of human mesenchymal stem cells derived from umbilical cord and bone marrow. Stem Cells. 2007;25(6):1384–92.
55. Barry F, Boynton RE, Liu B, Murphy JM. Chondrogenic differentiation of mesenchymal stem cells from bone marrow: differentiation-dependent gene expression of matrix components. Exp Cell Res. 2001;268(2):189–200.
56. Hernández-Bule M, Trillo M, Martínez-García M, Abilahoud C, Úbeda A. Chondrogenic differentiation of adipose-derived stem cells by radiofrequency electric stimulation. J Stem Cell Res Ther. 2017;7(407):2.
57. Abdi R, Fiorina P, Adra CN, Atkinson M, Sayegh MH. Immunomodulation by mesenchymal stem cells: a potential therapeutic strategy for type 1 diabetes. Diabetes. 2008;57(7):1759–67.
58. Shi M, Liu ZW, Wang FS. Immunomodulatory properties and therapeutic application of mesenchymal stem cells. Clin Exp Immunol. 2011;164(1):1–8.
59. Song L, Webb NE, Song Y, Tuan RS. Identification and functional analysis of candidate genes regulating mesenchymal stem cell self-renewal and multipotency. Stem Cells. 2006;24(7):1707–18.

Genomic and Proteomic Monitoring of Stem Cell-Derived Exosomes

Erdal Karaöz and Eda Sun

Abstract The mesenchymal stem cell (MSC) transplantations are increasing day by day in clinical therapies. Recent studies have described that the treatment activities of MSCs are mediated by paracrine factors secreted. These paracrine factors are spread through messenger clouds as exosomes secreted from MSCs, which are also involved in the cellular communication via transferring their mRNA, miRNA, and protein contents horizontally. With the understanding of this mechanism, cell-free therapies have begun to create a new path in MSC based therapies as a cell-free treatment. It is important to understand how much the mRNA, miRNA, and protein contents of the MSCs and their secreted exosomes are overlapping. Additionally, the content of these exosomes must be determined in order to both define their efficacy and to be able to select their most suitable sources. Therefore, in this chapter, we have described different transcriptomic and proteomic methods for displaying the contents of the exosomes.

Keywords Mesenchymal stem cell derive exosomes · Cell-free therapies · Transcriptome · Proteome

---

E. Karaöz (✉)
Center for Stem Cell and Tissue Engineering Research & Practice,
İstinye University, İstanbul, Turkey

Medical Faculty, Histology and Embryology Department,
İstinye University, İstanbul, Turkey

Center for Regenerative Medicine and Stem Cell Research and Manufacturing,
Liv Hospital, İstanbul, Turkey

E. Sun
Center for Stem Cell and Tissue Engineering Research & Practice,
İstinye University, İstanbul, Turkey

Medical Faculty, Histology and Embryology Department, İstinye University, İstanbul, Turkey

B. Arjmand (ed.), *Genomics, Proteomics, and Metabolomics*, Stem Cell Biology and Regenerative Medicine, https://doi.org/10.1007/978-3-030-27727-7_5

1 What Is Exosome?

The exosomes were first described by Johnston et al. in 1983 as vesicles that are responsible for the removal of cell debris [1]. Exosomes are bilipid membrane vesicles, containing different types of cargos including mRNAs, miRNAs, soluble proteins that are responsible for the cellular interaction, and transmembrane proteins presented in the appropriate and functional orientation [2–4]. These vesicles play an important role in intercellular communication, signal transduction, genetic material transfer, and regulation of the immune response [5].

The first report of a cellular function of exosomes was the shedding of the *transferrin* receptor that carries iron, maturing reticulocytes [1, 6]. So, it was understood that exosomes can communicate with target cells through several mechanisms.

2 Isolation of Exosomes

MSCs were cultured in a serum-free medium, in order to avoid the contamination of protein that will come from serum, to produce a concentrated and an adequate amount of exosome. Exosomes were produced in a humidified atmosphere containing 5% CO_2 at 37 °C for 24 h then the serum-free medium was collected and centrifuged at 300 g for 5 min for discarding the remaining cells until the last step. The pellet is thrown away and the supernatant is used for the following step. The second and the third step are designed to eliminate dead cells and cell debris by successive centrifugations at increasing speeds. At the second step, the supernatant was centrifuged at 1000 g for 10 min and in the third step, the supernatant was centrifuged at 5000 g for 20 min. The final supernatant is ultra-centrifuged at 100,000 g for 70 minutes. At last, the collected pellet is consisting of the small vesicles that correspond to exosomes from the pellet.

3 Characterization of Exosomes

After the isolation, exosome identities are determined by different types of methods that include atomic force microscopy (AFM), scanning electron microscopy (SEM), dynamic light scattering (DLS), flow cytometry (FCM), western blotting, nanoparticle tracking analysis (NTA), transmission electron microscopy (TEM), or ELISA. Characterization step is important to show that small particles are indeed exosomes.

As shown in Fig. 1, isolated exosomes were labeled with CD81, CD9, and CD63 and the characterization data were acquired with flow cytometry analysis. The most common markers used are tetraspanins such as CD9, CD63, and CD81.

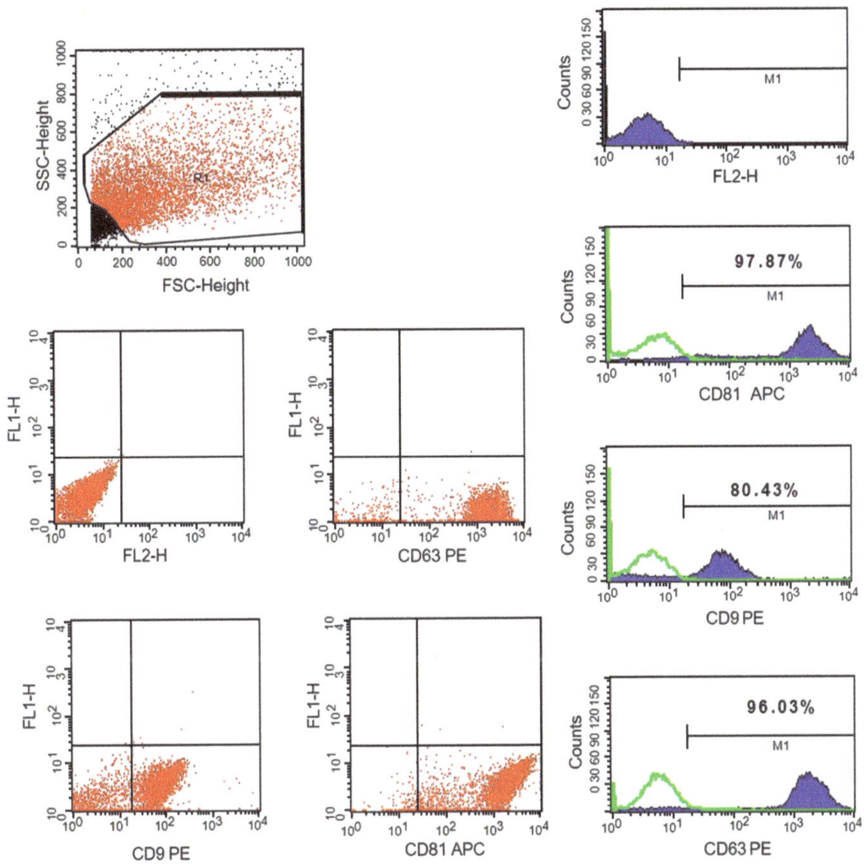

Fig. 1 Flow cytometry characterization analysis of cell-surface markers in WJ-MSC derived exosomes. MSC derived exosomes expressed markers including CD9, CD63, and CD81

4 Exosomes Derived from Mesenchymal Stem Cells

Mesenchymal stem cells (MSCs) are multipotent cells that are being clinically explored and used in regenerative medicine experimentally. In one of these experimental studies, it is demonstrated that MSCs significantly reduced the damaged myocardial ischemia area in pig and mouse models [7]. One of the important findings of this study was that the size of the active components was in the range of 50–200 nm as a result of fraction studies. Subsequent biophysical studies have characterized the biologically active components as exosomes. According to this study's results, the importance of exosomes was understood in MSC-based cellular therapies. By this study, it was understood that MSC-derived exosomes mimic the ability to affect the activity of immune cells (B, T, NK, dendritic cells, and macrophages), which are characteristics of MSCs [8]. MSC-derived exosomes have been shown to be immunologically active due to suppression of proliferation and IFN-y secretion by anti-CD3 and anti-CD28 stimulated T cells [9]. Another study demonstrated the immunoregulatory effect of exosomes that they enhance the survival after allogenic skin transplants in mice by increasing T cell polarization [10].

The current common opinion is the secretory factors of MSCs are regulating the inflammation, reprogramming the immune cells, and promoting the endogenous repair pathways [11]. Recent studies showed that MSCs produce extracellular circulators in various sizes, which are carrying the mRNA, microRNAs, and proteins and they make a horizontal transfer of these cargos to the target cell to induce non-autonomous changes as a therapeutic. These extracellular circulators are called exosomes.

Extracellular vesicles are able to affect cell phenotype, recruitment, proliferation, and differentiation in a paracrine manner. These paracrine effects of EV have a potential benefit in regenerative medicine by communicating with their target cells through cell-to-cell crosstalk. Firstly, transmembrane proteins on the EV membrane can interact with receptors on the cell membrane. These receptor-ligand interactions can then activate signaling cascades to affect target cells. EV can also fuse with their target cells to release their cargo, either by direct fusion with the cell membrane or by endocytosis, after which mRNAs, miRNAs, and proteins are released into the cytosol. Fusion of EV with target cells can either occur directly at the cell membrane or after endocytosis. After fusion, mRNAs transferred by EVs can be translated into protein, and delivered miRNAs inhibit mRNA translation and affect cellular processes. The cargo and function of EV depend on their producing cells, and it has been shown that also cellular stress affects EV content, suggesting that intercellular communication through EV is a dynamic system, adapting its "message" depending on the condition of the producing cells [2, 3, 12, 13].

5 The Roles of Exosomes in Regenerative Medicine

It is understood that the main actors of cellular therapies are the exosomes they secrete to the environment. As such, exosomes are important candidate biomolecules for use in regenerative medicine. The use of exosomes obtained from the correct source is very important to achieve the most effective result in clinical use. Therefore, it is important to determine the cargo contents of exosomes derived from different mesenchymal stem cell sources. This is only possible with proteomic and genomic studies.

MSCs derived exosomes have a wide range of cargo which can directly modulate the microenvironment through proteins localized on the exosome membrane or through the release of luminal proteins or RNA cargos. Tissue homeostasis is critical to proper and efficient tissue function and often disrupted in disease or injury. The survival of the tissue, recovery of normal tissue function, and initiation of repair are contingent on the timely restoration of tissue homeostasis. Delay of regeneration could lead to irreversible tissue damage. In this regard, MSC derived exosomes carrying a complex cargo of biologically active materials encapsulated within a receptor-studded lipid membrane represent the most efficient means to affect a timely restoration of tissue homeostasis. Hence, we postulate that MSCs exert their stromal support function through their exosomes to restore tissue homeostasis to diseased and injured tissues and to facilitate tissue repair and regeneration.

6 Monitoring of MSC Derived Exosomes

According to the information that rely on the genomic and proteomic databases, the MSC derived exosome contains at least 800 proteins and >150 microRNAs and mRNAs [14, 15]. For this reason, genomic and proteomic techniques have been used to determine the cargoes carried by exosomes relative to the cells they originate from, which is crucial for proper clinical use (Fig. 2).

6.1 Genomics Monitoring of MSC Derived Exosomes

6.1.1 Next Generation Sequencing

The New Generation Sequencing (NGS), which is the output of the Human Genome Project, has a history based on two-dimensional chromatography in the 1970s. In 1977, with the development of the Sanger chain termination method, the world of science has gained the ability to reproduce reliably and with the same accuracy. In 1988, the development of the "first generation" array products with capillary electrophoresis led to the production of discrete products of 84 kilobases (kb) 1

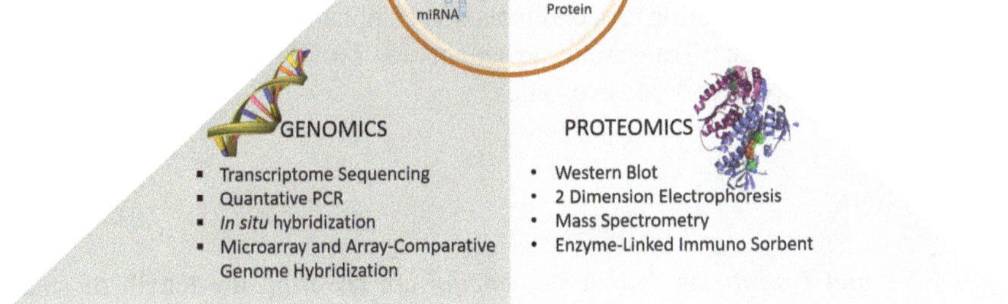

Fig. 2 Genomic and proteomic monitoring techniques of stem cell-derived exosomes

gigabase (GB) per run [16]. A short reading and parallel sequencing techniques have basically created a different approach. Because of these revolutionary qualities in genomic science, it is launched as "New Generation."

With this revolutionary development, when 1 gigabase (GB) of data is obtained as a result of single disposal in 2005, a surprising increase in 2014 and a single sequencing data about 1000 times more; 1.8 terabases reached (1GB=10^9 base pairs; 1 terabases=10^{12} base pairs). As a consequence of these advances, it is thought that thousands of people will be able to explore genomic data and discover the differences between genes with critical prescribing that cause diseases such as cancer, autism, heart disease, or schizophrenia [17].

The NGS field allows researchers to find answers to questions about the genome, transcriptome, or epigenome by a variety of methods. This wide variety of methods is basically different depending on whether the starting material is DNA or RNA. We can examine NGS technology under three main headings as genomic, transcriptomic, and epigenomics.

6.1.2 Transcriptome Sequencing

The primary library preparation for transcriptomic sequencing is RNA, and all technologies under this heading are also referred to as RNA-Recognition. Once the RNA library has been prepared in the RNA sequencing methodology, the ribosomes are removed and the cDNA is subsequently synthesized starting with a sample preparation protocol.

The considerable advantage of transcriptomic sequencing is that it is actually a gene expression study. Because of this feature, we present a snapshot of all transcriptome data. All transcriptome data provides a comprehensive examination of the cellular transcriptional profile. This technology provides the identification of alternative cloning sites, novel transcripts, and gene fusions that other NGS technologies cannot provide us.

Targeted RNA sequencing technology is a technology that focuses on genes that are members of a path clustered under a certain biological property and provides data on the subset of the advantages of transcriptomic sequencing technology that are determined at high resolution.

Transcriptome sequencing can reveal differences in the expression profiles of genes involved in regeneration especially exosomes obtained from different mesenchymal stem cell sources. Transcriptome sequencing provides the best coverage of mRNA and microRNAs in MSC exosomes.

6.1.3 miRNA

Small RNA and Untranslated RNA sequencing are covering the length of small RNAs that is 18–22 base pairs. Small RNAs, untranslated RNAs, and microRNAs are acting as regulators such as gene suppressor or gene silencing elements during gene expression. This technology, including in the transcriptomic group, offers transcriptional and translational regulatory effects of RNAs including this group.

6.1.4 Quantitative PCR

The quantitative real-time PCR method is a PCR method that can quantitatively obtain the result by measuring the fluorescence signal which increases simultaneously with the nucleic acid amplification. mRNA expression levels are generally determined by hybridization of the labeled probes to the mRNA itself or to the cDNA generated from the mRNA. As the target cDNA becomes double-stranded during the polymerization step following the binding of the primers, the amount of dye bound to the DNA increases and the resulting increase in the amount of fluorescence is observed. Relative gene expression levels obtained by PCR are calculated by the ΔCT method [18].

6.1.5 In Situ Hybridization

In situ hybridization (ISH) is a powerful technique that allows the display of nucleic acid sequences (DNA and RNA) and uses the principle of the double-stranded nucleic acid formation. The most important feature of this hybridization method is the recognition of nucleic acids in the exosomal environment. Detection of nucleic acids in the exosomal medium is important to determine both the gene expression and gene loci.

Fang et al. determined the microRNAs that play role in suppressing myoblast formation by transcriptome sequencing and then they showed miRNAs detected by sequencing in the exosomes by ISH [19].

6.1.6 Microarray and Array-Comparative Genome Hybridization (aCGH)

DNA microarrays contain thousands of fixed DNA sequences organized in a region with the most microscopic lamellar size. These microarrays are used in gene variants or in genotyping analysis or in gene expression analysis, enabling thousands of genetic analyses at the same time. The cellular sample used for genotyping is genomic DNA (gDNA). DNA containing exosomes are secreted by cancer stem cells and therefore genotyping can be done with exosomes obtained from cancer stem cells. For gene expression analysis, the mRNA molecule population of an exosome is converted to cDNA and labeled with a fluorescent dye. This mixture is then exposed to a gene chip, which is a glass slide or membrane containing thousands of small DNA regions, each of which corresponds to a different gene. The amount of fluorescence bound to each region corresponds to the ratio of mRNA in the sample.

Lou et al. discussed the biogenesis of MSC exosomes and their physiological functions and highlighted the specific biochemical potential of MSC-derived exosomes in restoring tissue homeostasis by discussing the outputs of microarray [20].

aCGH is a microarray-based system which is used to compare two different samples. The difference of this technique from a microarray is that the compressive samples can be stained with different fluorescent dyes. Thus, showing the differences in the gene expression profiles between the two samples.

6.2 *Proteomic Monitoring of Exosomes*

6.2.1 Western Blot

Western blotting is the most common method used to determine the specificity of a target protein in a complex mixture separated by molecular weights, weights and isoelectric points with different gel systems. Proteins separated in the gel system are transferred to the membrane without staining and then the target protein is detected by a specific immunological marker. In this detection method, the protein is treated with its own unique antibody. The regions in which the antigen-antibody linkage takes place are determined by means of a specific marker in the antigen.

6.2.2 2 Dimension Electrophoresis (2-DE)

After identifying or synthesizing a particular protein, it is necessary to separate and analyze it with 2-DE to determine whether it has undergone post-translational modification and what its function is.

2-DE is important for separating thousands of proteins in a mixture in one step. This method is in principle a combination of two separation techniques. The first step is the isoelectric focus, which is a pH gradient separation. The second step is the SDS-PAGE. The gel containing the proteins separated by isoelectric points is placed on a second gel and separated according to the molecular mass. The use of two separation techniques together provides a perfect separation. This method is necessary to use when separating mixtures containing a large number of proteins. In a recent report by Fraga et al., Wharton's jelly-derived exosomes were characterized by 2-DE [21].

6.2.3 Mass Spectrometry

In mass spectrometry, neutral molecules are ionized and positively charged ions are separated by mass/charge ratio as they pass through the electric or magnetic field. By using this method, Pires et al. determined protein expression profiles of exosomes derived from MSC of different origins [22]. They confirmed that the secretome of MSCs isolated from different tissue sources is rich in neuroregulatory molecules represents an important asset not only for the development of future neurodegenerative strategies but also for their use as a therapeutic option for human clinical trials.

6.2.4 Enzyme-Linked ImmunoSorbent Assay (ELISA)

ELISA is an immunological assay used to measure proteins and glycoproteins in exosomes. ELISA assays are generally carried out in 96-well plates, which need to be special absorbent plates to ensure the antibody or antigen sticks to the surface. Each ELISA measures specific antigen-anticore bindings. The advantages of ELISA is based on the use of specific antibodies, allowing higher sensitivity levels in the detection of antigens, and thus enabling the identification of proteins present in low concentrations [23]. Moreover, it provides an accurate quantification of proteins.

Jiang et al. have determined the TGF-β levels in fibrotic liver tissues treated with PBS or human umbilical cord MSC-Exo by ELISA so they showed that exosomes were inhibited the TGF-β. It means that exosomes may have potentials of oxidation resistance and anti-apoptosis in liver fibrosis [24].

7 Conclusion

Exosomes are extra-cellular vesicles that contain miRNAs, DNA or protein and participate in cell-to-cell communication and protein and RNA delivery. Owing to this communication and transfer feature, exosomes are very important therapeutics for cell-free treatment in regenerative medicine. Detection of the entire content of the information that exosomes send to the target cell will be possible by genomic and proteomic studies.

References

1. Pan BT, Johnstone RM. Fate of the transferrin receptor during maturation of sheep reticulocytes in vitro: selective externalization of the receptor. Cell. 1983;33:967–78.
2. Stoorvogel W, Kleijmeer MJ, Geuze HJ, Raposo G. The biogenesis and functions of exosomes. Traffic. 2002;3:321–30.
3. Théry C, Zitvogel L, Amigorena S. Exosomes: composition, biogenesis and function. Nat Rev Immunol. 2002;2:569–79.
4. Valadi H, et al. Exosome-mediated transfer of mRNAs and microRNAs is a novel mechanism of genetic exchange between cells. Nat Cell Biol. 2007;9:654–9.
5. Raposo G. B lymphocytes secrete antigen-presenting vesicles. J Exp Med. 1996;183:1161–72.
6. Harding C, Heuser J, Stahl P. Receptor-mediated endocytosis of transferrin and recycling of the transferrin receptor in rat reticulocytes. J Cell Biol. 1983;97:329–39.

7. Timmers L, et al. Reduction of myocardial infarct size by human mesenchymal stem cell conditioned medium. Stem Cell Res. 2007;1:129–37.
8. Burrello J, et al. Stem cell-derived extracellular vesicles and immune-modulation. Front Cell Dev Biol. 2016;4:83.
9. Blazquez R, et al. Immunomodulatory potential of human adipose mesenchymal stem cells derived exosomes on in vitro stimulated T cells. Front Immunol. 2014;5:556.
10. Zhang B, et al. Mesenchymal stem cells secrete immunologically active exosomes. Stem Cells Dev. 2014;23:1233–44.
11. Phinney DG, Pittenger MF. Concise review: MSC-derived exosomes for cell-free therapy. Stem Cells. 2017;35:851–8.
12. Pegtel DM, et al. Functional delivery of viral miRNAs via exosomes. Proc Natl Acad Sci. 2010;107:6328–33.
13. Johnstone RM, Mathew A, Mason AB, Teng K. Exosome formation during maturation of mammalian and avian reticulocytes: evidence that exosome release is a major route for externalization of obsolete membrane proteins. J Cell Physiol. 1991;147:27–36.
14. Lai RC, et al. Proteolytic potential of the MSC exosome proteome: implications for an exosome-mediated delivery of therapeutic proteasome. Int J Proteom. 2012;2012:1–14.
15. Chen TS, et al. Mesenchymal stem cell secretes microparticles enriched in pre-microRNAs. Nucleic Acids Res. 2009;38:215–24.
16. Illumina. An introduction to next-generation sequencing technology.
17. Fallows J. When will genomics cure cancer? The Atlantic.
18. Livak KJ, Schmittgen TD. Analysis of relative gene expression data using real-time quantitative PCR and the 2(-Delta Delta C(T)) method. Methods. 2001;25:402–8.
19. Fang S, et al. Umbilical cord-derived mesenchymal stem cell-derived exosomal MicroRNAs suppress myofibroblast differentiation by inhibiting the transforming growth factor-β/SMAD2 pathway during wound healing. Stem Cells Transl Med. 2016;5:1425–39.
20. Lou G, Chen Z, Zheng M, Liu Y. Mesenchymal stem cell-derived exosomes as a new therapeutic strategy for liver diseases. Exp Mol Med. 2017;49:e346.
21. Fraga JS, et al. Unveiling the effects of the secretome of mesenchymal progenitors from the umbilical cord in different neuronal cell populations. Biochimie. 2013;95:2297–303.
22. Pires AO, et al. Unveiling the differences of secretome of human bone marrow mesenchymal stem cells, adipose tissue derived stem cells and human umbilical cord perivascular cells: a proteomic analysis. Stem Cells Dev. 2016;25:1073–83.
23. Kupcova Skalnikova H. Proteomic techniques for characterization of mesenchymal stem cell secretome. Biochimie. 2013;95:2196–211.
24. Jiang W, et al. Human umbilical cord MSC-derived exosomes suppress the development of CCl_4-induced liver injury through antioxidant effect. Stem Cells Int. 2018;2018:1–11.

Proteomics Approaches Applied to Regenerative Medicine: Perspectives in Stem Cell Proteomics

Saeed Heidari-Keshel, Mostafa Rezaei-Tavirani, Azam Rahimi, Farshid Sefat, and Arash Khojasteh

Abstract Stem cells are able to maintain self-renewal for a long time. Under certain conditions, these cells may differentiate into adult and functional cell types. According to their origin and developmental capacity two essential types of stem cells are classified: embryonic and adult or mesenchymal stem cells. Human embryonic stem cells (hESCs) are able to generate adult somatic tissue cells because they are pluripotent cells but other types of stem cells are called tissue-specific because each of them is particular to specific tissue. Proteomics studies have the potency to define molecules and pathways pivotal for cell biology and the strategies by which cells can participate in transplantation and repair. The proteome is the whole protein content that expressed in an organism. The proteomics final goal is to characterize information flow, with protein analysis, pathways, and networks. In this chapter, we tried to explain proteomics studies of embryonic stem cells and discuss their extraordinary characteristics or properties which make them interesting in medical investigations. The hESCs differentiation pattern prepares a model for examining the cellular and molecular events of early development that is cellular and molecular with a significant potency for the progression of proteome analysis. Proteins expressed by a large number of genes are characterized and the pattern of expression has been compared between iPSCs and ESCs. We will investigate adult or mesenchymal stem cell proteomics and after that about the proteomic study of iPSC and finally the methodology used for both of them.

Keywords Stem cell · Proteomics · Regenerative medicine

S. Heidari-Keshel · A. Rahimi · A. Khojasteh
Department of Tissue Engineering and Applied Cell Sciences, School of Advanced Technologies in Medicine, Shahid Beheshti University of Medical Sciences, Tehran, Iran

M. Rezaei-Tavirani (✉)
Proteomics Research Center, Shahid Beheshti University of Medical Sciences, Tehran, Iran

F. Sefat
Biomedical and Electrical Engineering Department, School of Engineering, University of Bradford, Bradford, UK

Interdisciplinary Research Centre in Polymer Science and Technology (IRC Polymer), University of Bradford, Bradford, UK

B. Arjmand (ed.), *Genomics, Proteomics, and Metabolomics*, Stem Cell Biology and Regenerative Medicine, https://doi.org/10.1007/978-3-030-27727-7_6

1 Introduction

Regenerative medicine is recently considered one of the most promising therapeutic strategies. Reducing functional cell degeneration with regenerating tissue cells is the golden aim of this method. Stem cells are self-renewal under proper conditions with mitotic cell division and can generate, maintain, and capable of regenerating damaged tissue by producing major cell types [1].

Stem cells have the ability to remain self-renewing and proliferating. These cell types are able to generate various adult and functional cell types, under specific conditions. Depending on their origin and differentiation abilities, stem cells are divided into two principal types: embryonic stem cells and adult stem cells. Adult or mesenchymal stem cells are mainly known as the cells that are particular to their specific tissues while embryonic stem cells (ESCs) can differentiate into adult somatic cells as they are pluripotent stem cells. The origin of human tissue-specific stem cells is not embryonic tissues; therefore, these cells can produce particular cells of their origin tissue. These cells are an extremely favorable tool to understand the development and are desirable for regenerative medicine. Hence, conducting methods for the differentiation of these cells and chose these for analysis and specific applications are an area of interest in a numerous investigation worldwide [2] (Fig. 1).

Proteomics studies of specific cells can define molecules and pathways essential for cell biology and the methods by which cells can take part in transplantation and repair [3]. The important role of biological molecules such as RNA, DNA, protein, and peptide for life support has been emphasized from the beginning of the biological science development [4]. The proteome is defined as all proteins content expressed in a cell or organism. One of the proteome features is being dynamic

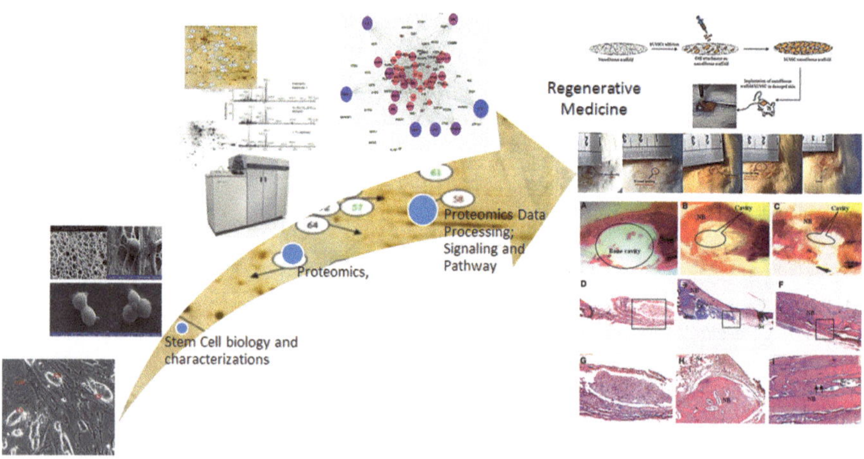

Fig. 1 The proteomics approach for advancement of regenerative medicine and for human disease diagnostics and treatment

because in response to development and aging process the expression and post-translational modifications may vary. The final aim of proteomics is to characterize the cells or organism information by analyzing proteins, pathways, and networks [5].

Metabolomics is determined as the science responsible for the study of whole metabolite indexes in tissue components under appropriate conditions and the metabolome concept is essentially a metabolism result. The metabolome also consists of other components absorbed from the environment. Eventually, it is desirable to use the information obtained from a combination of different omic science methods to expand treatment and prevention methods [6].

Metabolomics complete other omics techniques such as transcriptomics and genomics to present the protein's exploration in a cell or organism, and to realize a specific protein structure and function. The technologies which based on proteomics are employed in a variety of potentials for various experiments such as vaccine production, finding diagnosing markers, understanding the mechanism of pathogenesis, change in the protein expression patterns in response to various elements, and explanation of protein pathways in various diseases [7].

The conventional procedures for peptides and proteins isolation and purification are chromatography-based methods mostly like size exclusion (SE), ion exchange (IE), and affinity chromatography [8–10].

For specific proteins analysis, western blotting and enzyme-linked immunosorbent assay (ELISA) are generally used. The methods mentioned additionally are limited to a few individual proteins analysis, however, unable to explain the expression level of proteins [11, 12].

In 2-D gel electrophoresis, isoelectric focusing is the first step in which samples are determined according to their charge, and in the second step based on molecular weight. The result includes lots of spots in different sizes, each of them demonstrates a specific protein. This technique is mostly applied to compare two protein samples to discover particular protein diversity [13, 14].

To understand the three-dimensional structure of proteins and also to compare their structure understanding structural proteomics is desirable. It gives voluminous data about the shape, function, and structure of protein collections present in particular cells or organisms. NMR spectroscopy and X-ray crystallography are different technologies mainly used for mass spectrometry that is a structure specification. This technique is based on analysis and generates spectra of the masses containing a sample of protein [15, 16]. Chemical ions that generate charged molecules and evaluating their charge ratios are the bases of mass spectrometry.

One of the most favorable techniques for recognition of certain proteins is MALDI-TOF. This proteomics technique is quantitative, gel-free, and based on chemical labeling. To compare two proteomes quantitatively, a sample protein will label with a heavy isotope and another one label with the isotopically light version. On the other hand, each sample tagged with isotope-coded reagents. The maintained peptides are inspected by LC-MS. This method usually applied the quantification of the dependent proteins in several samples. Visible isotope-coded affinity tags are other techniques in ICAT visible label which lets the proteins electrophoresis sites to be simply observed [17]. Another gel-free method performed to quantify sample proteins is iTRAQ that is isobaric tags for absolute and relative quantitation. This technique applied to study quantitative modifications in whole protein content. Tagging the amines of a side chain of peptides and N-terminus of it avoided peptides from digestions via modifying mass tags, 4-plex and 8-plex that are the reagents capable of labeling every peptide from various samples. The samples can be analyzed with mass spectrometry [18].

In this study the impact of proteomics in stem cell research especially embryonic stem cells (ESCs), adult stem cells (ASCs), and induced pluripotent stem cells (iPSCs) is discussed. It is proposed that the reviewed subjects can be a useful guide to apply in regenerative medicine.

2 Proteomics and Stem Cells

Each type of stem cell is characterized by two specific features. Unrestricted self-renewing, the procedure by which stem cells divide to make more stem cells, immortalizing the stem cell as a valuable reserve throughout the life, and maintenance of the undifferentiated state is the first feature. The second is the potency to generate various and differentiated cell types, in contrast to general somatic cells which committed to unique lineage. Tissue-specific or adult stem cells may discover within the majority of mature organisms tissues and are considered to operate tissue survival and regeneration in long term [19].

The possibility of stem cells characterizing at a proteome-wide scale was considerably elevated by the developments in MS-based methods. As a result, an associated development in the proteomic resolution number of cells is detected. MS-based proteomic techniques are capable to investigate stem cells by presenting samples with non-stem cells [20]. About 10 years ago, only a limited number of stem cell studies applied proteomic except research using antibodies. Recently, the importance of integrating data of proteomics and transcriptomics has been emphasized to gain a better view of stem cell activities [21]. More pharmaceutical perspective was given by Klemm et al. to highlight the potential of the proteomic methods to indicate neurotoxic factors which active compositions may have on human embryos instead of animal models, by using hESCs [22]. All available proteomics studies were discussed by Roche et al. and used them to characterize MSCs [23]. Park and his colleagues assembled a complicated list of proteins derived from MSCs and suggested them to use it as a reference pattern for MSCs proteome [24]. These adult progenitors knew as a possible subjects for tissue-engineering and differentiated cell replacement; this is primarily because they are autologous cells and capable to induce immunological responses. A comprehensive type of the analytical method was published by Baharvand and his co-workers to profile stem cell proteomes [25]. The information about cell application, passage rate, differentiation potency, the approach of proteomic and number of protein characterizations are evaluating attractively. Recently, proteomic characteristics of human ESCs were evaluated in detail by Van Hoof [26]. The availability of hESCs proteome datasets is limited to studies [27], restricting the comparison between mESCs and different tissue-specific cell of the same species. Several hESCs proteomics studies have been done and their obtained data about differentiated generations have been published. Combining advanced data in favorable technologies is used to evaluate the proteome, transcriptome, and genome of neural, embryonic, and hematopoietic stem cells [28]. The ESCs characterization at the epigenome, proteome, and transcriptome degree exhausted with Stanton and Baker in a succinct overview [29]. In conclusion, the term proteomics refers to various techniques that can identify proteomes and has been widened to include methods capable of characterizing proteins and both their functions and structures analyzing at genome stage. Proteomics contains various methods for characterizing interactions between proteins, protein chips which are antibody-based for identifying peptides/proteins and making crystallography screens to prepare structural analysis [30].

2.1 *Proteomic Studies of Embryonic Stem Cells (ESCs)*

Human embryonic stem cells or hESCs with significant biological characteristics have properties that help them to be marvelous in regeneration research. They show a significant need for proteome analysis progression. These cells supply a great model system to investigate differentiation process through embryonic development [31]. The hESCs are isolated from the blastocyst inner cell mass, and they have nearly unlimited self-renewal, besides differentiation potency into all cell types existing in three embryonic germ layers. The embryonic stem cells differentiation in vitro provides an examining model for the early development mechanisms that may be cellular or molecular [32].

ESCs characterization and molecular analysis are critical and accomplished to ensure that we can purify these cells as hESC applications are an important tool of regenerative medicine [33]. We can correlate the cellular alteration to behold therapeutic effects. Cell surface proteins are membrane markers used for isolation and sorting because they are the most significant branch of SC proteins [34]. Also, these proteins are capable to work as anchors and receptors for extracellular molecules and membrane structures. Extrinsic proteins exist on the surface of the cell, regulating arrival and efflux current of various intra and extracellular signaling molecules. Our information about surface marker is confined due to the difficulties associated with their crystallization and cloning [35]. Using the ability of self-renewal, different stem cells can differentiate into all cell types. According to these characteristics, various stem cells used in both developmental studies and regenerative medicine. Therefore, it is critical to know the controlling mechanism of pluripotency maintenance and self-renewal of these cells. By changing in energy metabolism, several metabolic pathways control the pluripotency and renewal maintenance of these cells [36]. These cells are responsible for regeneration in tissue engineering applications, however, before this becomes a reality various hurdles need to be overcome. Distinct explanation of the factors that are necessary to maintain, being self-renew and pluripotent are the properties of these cells [37]. Selecting and isolating precursor cells and differentiating them with particular abilities have the potential to hold the key to totally exploiting them and techniques for differentiation for clinical application. Several challenges currently prevent the application of subpopulations of stem cells. First, the heterogeneity that stands among subpopulations makes us unable to characterize the cells phenotypically, isolate and identify them from differentiating progeny. Also, in spite of a rapid improvement in stem cell biology, presently, researchers rely on indirect techniques to establish characteristics and functions of stem cells. These challenges are related to basic and translational research usage [38].

Innovative methods have been increased to overcome the challenges in surface protein identification. Generally, two sets of methods have developed, that have investigated an ability to characterize the surface markers more desirable than the entire cellular proteome analysis and thought to help characterize stem cell surface markers. These consist of "physical enrichment" and "affinity enrichment" of plasma membranes associated proteins. Differential centrifugation has been done for enrichment strategies and membrane proteins [39].

Identifying the factors that contribute to self-renewal is beneficial in controlling teratoma formation after transplantation and signaling pathways activation in consecutive differentiation. ESCs characterization and wide adjustment of the various lines of ESCs presented a perspective of ESC renewing and the pathways involved in differentiation. However, the proteomics that based on mass spectrometry extended to characterize biological processes at stages that rival those done with microarray and other strategies [40]. It is achievable to study and detect several types of post-translational modifications, such as phosphorylation, ubiquitination, glycosylation, methylation, and acetylation of histones that involve the structure and function of the protein [41–43].

The proteomics based on mass spectrometry may identify any protein with a discovery-oriented strategy. On the other hand, mass spectrometry is not limited to the high-throughput characterizing huge samples of proteins [44]. A proteomic examination normally follows the similar stages: (1) prepare the samples, such as isolation of peptide or protein, (2) mass spectrometry analysis and quantitation and characterization of proteins, and (3) annotation these identifications functionally [45–48].It is clear that ESCs have significant differentiation and regeneration capacity and provide insight into the developmental process, drug screening, and toxicity research. MS has evolved into a useful tool that interestingly suited to the characterization of subcellular and cellular proteins [49]. The extended mass spectrometry tool generates proteome profiles consisting of several hundred to thousands of proteins and validating and identifying the roles of these candidates are not possible in vitro. Proteins identified in HES-2 hESCs or HUES-7 hESCs were isolated which include prominent 16 plasma membrane and 32 intracellular proteins. Because these proteins are expressed in undifferentiated cells, they all are likely to have a critical responsibility in hESCs, yet each of them needs validation with cellular and molecular biology methods. Remarkably, some proteins that belong to undifferentiated cells like hESC have known as critical for self-renewal process. Some of these proteins such as Sox2 [50] have been known in the later progeny that become differentiated. Furthermore, the expression phase of Oct4 (a growth factor) is regulated and a twofold addition in ESC expression effects in primitive mesoderm and endoderm differentiating process [51–53]. It seems that stemness depends on protein expression, on the location, and the amount of the ESCs protein. Heavy isotopes usage will permit quantitation of absolute and relative protein levels in ESCs to conduce to differentiate under various conditions. The strategies that used frequently rely on isotope tagging was demonstrated in cell culture and hardly precise when limit numbers of hESCs are applied [54, 55].

Using mass spectrometry-based methods, different candidates that may have a significant role in ESC self-renewing have been identified in tissue culture media by feeder layer to support generations of hESCs in the undifferentiated stage. Some of these were marked up as intracellular proteins that are ready to secrete [56]. Only a few have been tested, and none of them has proven to maintain the undifferentiated hESC mood or prevent differentiation enough over a long culture period. The multiple frameworks proteomics technics must be used to identify the valuable secretion proteins in hESC self-renewal pathway. Although many important proteins in the process of embryonic stem cell self-renewal are part of the secreted proteins from the feeder layer, it is possible that some of the serum proteins also be involved in the ESCs development and cultivation under non-differentiated conditions. Albumin and plenty of other proteins are working as a barrier against analyzing proteome and detecting the serum essential factors in. On the other hand, by reducing human serum factors decrease, the xeno-contamination risk and its origin are limited. Dynamics in intracellular trafficking, protein degradation, and expression is anticipated to generate a wealth of information preparing insight into hESCs pathways [57, 58]. Phosphorylated intracellular proteins identified in cells may investigate the ability to examine cell behavior in the expression of cellular proteins. An alteration in the transcription of genes has a distinct effect on the production of mRNA, stability, translation, degradation, PTMs, subcellular localization, and the rate of translation which all regulate the activity of proteins. In this respect, mRNA profile changes cannot show the true protein activity dynamics or represent a variety of signaling pathways happens inside the cells. Complete analyses of proteome and transcriptome have distinct complementary characters and in general, they produced due to the conflict of information sets, so they have been considered as two separate disciplines [59].

Mass spectrometry is a significant analysis and identity technology for peptide and protein samples characterization. MS high-quality spectra enable sampling of several proteins from one injection. Sequence ions may be produced by peptide ion segmentation in a mass spectrometry collision cell [60]. Proteomics that based on mass spectrometry is typically performed in an approach called "bottom-up," in which an enzymatical protein digestion to previous peptides happen to perform fractionation as well as characterization.

New capabilities of advance tools and software lead to the expansion of emergence technics called "top-down," in which complete proteins are analyzed [61, 62]. Bottom-up experiment detection depends on interpretation software that makes comparisons between peptide [63, 64].

MS may discover only a finite amount of ions in limited time, even with development in instrument scan speed. For the spread, injected sample components across a greater period, the chromatographic technique for isolation is used before mass spectrometry. The most prevalent technique is dissociation with a column called reversed-phase in which hydrophobic properties of the sample elements have a significant effect on partitions. It may be prevalent for an investigator to use a secondary fractionation technique before reversed-phase, for instance, an analytical methods like 1-D SDS-PAGE that is based on cation exchange are complementary and may develop proteome coverage [65, 66]. For increasing the biological importance of our information, quantitative methods of proteomic are often performed to compare cellular moods or conditions.

Mass spectrometers are not completely quantitative and very stable isotopic labeling is mostly used; however, label-free strategies are becoming popular [67]. One of these labeling techniques, named "stable isotope tagging with amino acids in cell culture," contains incorporation of culture media amino acids with a cell's proteome metabolically [68, 69]. In iTRAQ strategies, proteins should be labeled with tags following proteolytic digestion [70].

Tagged proteins are mixed at a determined ratio, and the sample abundance may be compared to rely on the specific peptide label amount that exists in the MS spectrum. Advancement in technology in peptide and protein fractionation methods, mass spectrometers, and quantitive technique has greatly increased the capability to analyze and adopt biological information from complicate cellular samples. According to the current development in the proliferate potency of hESCs in proteomic experiments, there is an enhancement in investigations of this significant cell type. Recent experiments rely on proteomics have increased our comprehension of hESC generation and potency with hESC investigation in the microenvironment, epigenetic control, membrane protein expression, and PTM state. The current development of proteomics methods made this science a more important approach in our conception of ESCs. Surely, studies of post-translational modifications may increase our conception of embryonic stem cells system function. Proteomics may have a critical function in the microRNAs emerging field, wherein quantitative technique may apply to model regulation processes up and down. On the other hand, ESCs proteomic analysis may integrate the large genomic data library of hESC [71].

2.2 *Proteomic Studies of Adult Stem Cells*

Adult stem cells as well as mesenchymal stem cells have more developmental ability and more commonly noticed to be unipotent to multipotent. MSCs typically generate only related family progeny, supply "new" cells to replace and particularize adult cells which need generation or repair. Hematopoietic stem cells (HSCs) are the prototype of restricted differentiation which produces all of the lymphoid and myeloid cell types [72]. MSC was first separated from bone marrow but certainly inhabit perivascular tissues like tendon, muscle, skin, adipose tissue, dental pulp, and placenta. Different population monitor exclusive phenotype and potency [73]. Generally, MSC will differentiate into mesoderm-derived tissues like chondrocytes, adipocytes, and osteocytes [74]. Plenty of studies have demonstrated transdifferentiation into endodermal or ectodermal cells and even differentiation into myocytes and cardiomyocytes

can develop in vitro but immunomodulation potential, proliferation, and differentiation ability of these cells are gradually limiting over time [75]. Currently, embryonic stem cell-derived MSC (ESC-MSC) was identified and has extreme proliferation ability and an increased cell source for regeneration. These cells consist of general markers of mesenchymal stem cells and monitor the same ability as bone marrow MSCs, such as immunomodulation, homing and marvelous regeneration, and differentiation potency. Cells will be arranged as MSC if they carry the specific surface protein markers such as CD105, CD90, and CD73, do not express hematopoietic specific markers such as CD45, CD19, and CD14, and show the potency to generate fat, cartilage, and bone. To develop mesenchymal stem cell-based therapy, a univocal MSC explanation is favorable molecular identification, as delivered by "Omics" strategies, is an attractive approach towards that aim. The newest aim is the usage of iPSCs to create individualized MSC at large scale before treatment. Creating molecular equivalency between MSCs with the various sources is a gold question for further ESC-MSC clinical usage. A wide proteomic and transcriptomic ESC-MSC identification and their checking to BM-MSC as well as MSCs were shown. The rationale for choosing this strategy relied on its adaptability to clinical applications, simplicity, and robustness. Two profiling methods were mixed for a wide identification: quantitative high-resolution nano-LC-MS/MS RNA, as well as deep sequencing, relied on stable isotope tagging with the amino acid in cell culture (SILAC) [76].

Plenty of proteomic attempts from ASCs have to operate 2-DE as a front-end fragmentation strategy former to MS identification. Merged datasets from proteomic profiling investigations define that the most complete collections of proteins in ASCs may be associated with metabolism [77].

The growth iTRAQ strategy that is a sensitive methodology composed of MS has let adjustment of purified HSCs and their progenitor with as few as 1 × 106 input cells. Attractively, studies output offers that HSCs are adapted for anaerobic environments [78]. iTRAQ is effective in defining unwell characterized hematopoietic progenitor cells as mainly erythroid. In this field 2-dimensional liquid chromatography or LC/LC fragmentation followed by tandem MS/MS utilizing has shown that osteogenic differentiation results from the centralizing of gene expression in a functional group rather than simply from the increased expression of new genes [79].

Significantly, it was mentioned that post-translational modifications can affect ASCs fate decisions. Surely, a phosphoproteomics, simplified with SILAC methodology, applied to study the effect of various factors and signaling on the differentiation of MSCs. Certainly, it was found that tyrosine phosphorylation mechanism moderates EGF and PDGF to impacted differentiation of MSC [80].

Characterization of protein transcription and interactome regulatory network in ESCs has given a new meaning in controlling development processes and the pluripotency of ASCs [81, 82]. Altogether, the research of ASCs proteomic profiling attained the fact that these strategies may guide to novel vision into the fundamental biology of the cells that are not mentioned using other means [83].

2.3 Proteomic Study of iPSC

Human-induced pluripotent stem cells or iPSCs recently were produced by retroviral transduction. These cells have significant potential for clinical research like disease modeling. Comprehensive research containing genomic and transcriptomic methods has been performed to identify iPSCs, but the molecular mechanisms remain insufficiently understood to recognize bona fide iPSCs from incomplete reprogrammed cells. Recently, distinct iPSC proteome analyses have been carried out to compare iPSCs and ESCs proteomes [84]. Proteins expressed by thousands of genes were characterized and the expression profiles have been compared between iPSCs and ESCs. However, the essential sample amounts and the total LC-MS/MS evaluation time are partly large, due to the need for pre-fractionation, transcriptomics approaches are evaluated. Nano-LC-MS/MS systems widely have been applied for proteomic investigations in several studies. It is considered that the iPSC can indicate the pathology of the disease better. The pluripotent nature of iPSCs may prepare an unlimited reserve of cells for several studies such as drug screening and cell therapy. It will be also important to generate transgene-free reprogrammed iPSC under GMP (Good Manufacturing Practice) conditions, for therapeutic application [85].

These stem cells (iPSCs) are important research instruments and capable to be a remarkable autologous cell source derived from iPSC for clinical applications. Before therapeutic application proper characterization of iPSC is required. These cells have derived from various cell types such as intestine epithelial cells. Human iPSC molecular characterization has been done recently on different biological levels, such as epigenetic evaluation, gene expression profiling, genomic DNA alterations, miRNAs functions in pluripotency, and alterations in genomic DNA. Up to now, quantitative proteomic methods have not been applied in the systematic characterization of human iPSC, and the proteome molecular variety of hiPSC from various somatic sources has not been directed. MS-proteomic methods on human iPSC presented previously vary remarkably, which complicates direct adjustment of these studies. Qualitative and quantitative adjustment of human iPSC and human ESC proteomes with that of adult cells enables the development of proteins for characterizing human iPSC that was examined in eight more human iPSC lines of various sources with different reprogramming methods. There is a list of proteins with a different expression between human iPSC isolated from fibroblasts. Eighty-nine proteins were generally down-regulated and 424 proteins were in general up-regulated in human iPSC (SB5-MP1) evaluate with fibroblasts [86].

3 Conclusion

Currently, stem cell is one of the important and most dynamic fields of the regenerative medicine relative research, for the reason that of the incomparable SCs properties such as differentiation potency and renewing that lead to all cells derived from ectoderm, mesoderm, and endoderm of the body. The iPSCs generation from autologous adult cells and the cultivation of cancer stem cells is one of the most promising titles in modern biology. Proteomics provides a great tool for deeply and extensively evaluating fundamental aspects of the biological events associated with stem cells and regenerative medicine such as pluripotency, differentiation, and treatment monitoring. Moreover, monitoring of these properties in vitro and in vivo would offer remarkable potential to develop cures of diseases that cannot be treated in the present day, particularly in the field of cell-based therapy. Since our comprehension of most of the cellular actions such as cell/stem cell self-renewal and differentiation of 2-dimensional and 3-dimensional cell culture is still very limited, it will be required to develop our science particularly on the level of protein.

References

1. Kingham E, Oreffo R. Embryonic and induced pluripotent stem cells: understanding, creating and exploiting the nano-niche for regenerative medicine. ACS Nano. 2013;3:1867–81.
2. Bernstein H, Hyun W. Strategies for enrichment and selection of stem cell-derived tissue precursors. Stem Cell Res Ther. 2012;3:1–14.
3. Lorelei D, Harley I. Neural stem cells (NSC) and proteomics. Cell Proteom. 2016;15:344–54.
4. Aebersold R, Mann M. Mass spectrometry-based proteomics. Nature. 2003;422:198–207.
5. Vlahou A, Fountoulakis M. Proteomic approaches in the search for disease biomarkers. Biomed Life Sci. 2005;814:11–9.
6. Atkinson S, Armstrong L. Epigenetics in embryonic stem cells: regulation of pluripotency and differentiation. Cell Tissue Res. 2008;331:23–9.
7. Aslam B, Basit M, Nisar M, Khurshid M, Rasool M. Proteomics: technologies and their applications. J Chromatogr Sci. 2017;2:182–96.
8. Jungbauer A, Hahn R. Ion-exchange chromatography. Methods Enzymol. 2009;463:349–71.
9. Voedisch B, Thie H. Size exclusion chromatography. In: Kontermann R, Dübel S, editors. Antibody engineering. Berlin: Springer; 2010. p. 607–12.
10. Hage DS, Anguizola JA, Bi C, Li R, Matsuda R, Papastavros E. Pharmaceutical and biomedical applications of affinity chromatography: recent trends and developments. J Pharm Biomed Anal. 2012;69:93–105.
11. Lequin RM. Enzyme immunoassay (EIA)/enzyme-linked immunosorbent assay (ELISA). Clin Chem. 2005;51:2415–8.
12. Kurien B, Scofield R. Western blotting. Methods. 2006;38:283–93.
13. Issaq H, Veenstra T. Two-dimensional polyacrylamide gel electrophoresis (2D-PAGE): advances and perspectives. BioTechniques. 2008;44:697–700.
14. Marouga R, David S, Hawkins E. The development of the DIGE system: 2D fluorescence difference gel analysis technology. Anal Bioanal Chem. 2005;382:669–78.
15. Ong S, Mann M. Stable isotope labeling by amino acids in cell culture for quantitative proteomics. Methods Mol Biol. 2007;359:37–52.

16. Wiese S, Reidegeld KA, Meyer HE, Warscheid B. Protein labeling by iTRAQ: a new tool for quantitative mass spectrometry in proteome research. Proteomics. 2007;7:340–50.
17. Kroksveen AC, Jaffe JD, Aasebo E, Barsnes H, Bjorlykke Y. Quantitative proteomics suggests a decrease in the secretogranin-1 cerebrospinal fluid levels during the disease course of multiple sclerosis. Proteomics. 2015;15:3361–9.
18. Perez-Riverol Y, Alpi E, Wang R, Hermjakob H, Vizcaino JA. Making proteomics data accessible and reusable: current state of proteomics databases and repositories. Proteomics. 2015;15:930–49.
19. Cox J, Mann M. Is proteomics the new genomics. Cell. 2007;130:395–8.
20. Cravatt BF, Simon GM, Yates JR. The biological impact of mass-spectrometry-based proteomics. Nature. 2007;450:991–1000.
21. Dai B, Rasmussen TP. Global epiproteomic signatures distinguish embryonic stem cells from differentiated cells. Stem Cells. 2007;25:2567–74.
22. Elling U, Klasen C, Eisenberger T, Anlag K, Treier M. Murine inner cell mass-derived lineages depend on Sall4 function. Proc Natl Acad Sci U S A. 2006;103:16319–24.
23. Freitas MA, Sklenar AR, Parthun MR. Application of mass spectrometry to the identification and quantification of histone post-translational modifications. J Cell Biochem. 2004;92:691–700.
24. Ginis I, Luo Y, Miura T, Thies S, Brandenberger R, Gerecht-Nir S, Amit M, Hoke A, Carpenter MK, Itskovitz-Eldor J, Rao MS. Differences between human and mouse embryonic stem cells. Dev Biol. 2004;269:360–80.
25. Harvey DJ. Proteomic analysis of glycosylation: structural determination of N- and O-linked glycans by mass spectrometry. Proteomics. 2005;2:87–101.
26. Ivanova N, Dobrin R, Lu R, Kotenko I, Levorse J, DeCoste C, Schafer X, Lun Y, Lemischka IR. Dissecting self-renewal in stem cells with RNA interference. Nature. 2006;442:533–8.
27. Jensen ON. Interpreting the protein language using proteomics. Nat Rev Mol Cell Biol. 2006;7:391–403.
28. Joephson R, Ording CJ, Liu Y, Shin S, Lakshmipathy U, Toumadje A, Love B, Chesnut JD, Andrews PW, Rao MS, Auerbach JM. Qualification of embryonal carcinoma 2102Ep as a reference for human embryonic stem cell research. Stem Cells. 2007;25:437–46.
29. Kirkpatrick DS, Denison C, Gygi SP. Weighing in on ubiquitin: the expanding role of mass-spectrometry-based proteomics. Nat Cell Biol. 2005;7:750–7.
30. Klemm M, Schrattenholz A. Neurotoxicity of active compounds establishment of hESC-lines and proteomics technologies for human embryo and neurotoxicity screening and biomarker identification. ALTEX. 2004;21(Suppl 3):41–8.
31. Trounson A, McDonald C. Stem cell therapies in clinical trials: progress and challenges. Cell Stem Cell. 2015;17:11–22.
32. Darabi R, Perlingeiro RC. Lineage-specific reprogramming as a strategy for cell therapy. Cell Cycle. 2008;7:1732–7.
33. Hassani SN, Totonchi M, Gourabi H. Signaling roadmap modulating naive and primed pluripotency. Stem Cells Dev. 2014;23:193–208.
34. Hughes C, Radan L, Wing Y, William L, Dean H, Gilles A. Mass spectrometry-based proteomic analysis of the matrix microenvironment in pluripotent stem cell culture. Mol Cell Proteomics. 2012;11:1924–36.
35. Shekari F, Han CL, Lee J, Mirzaei M, Gupta V, Haynes PA, Lee B, Baharvand H, Chen YJ, Hosseini SG. Surface markers of human embryonic stem cells: a meta-analysis of membrane proteomics reports. Expert Rev Proteomics. 2018;55:1–21.
36. Taleahmad S, Hassani SN, Hosseini Salekdeh G, Baharvand H. Metabolic signature of pluripotent stem cells. Cell J. 2018;20:388–95.
37. Soteriou D, Iskender B, Byron A, Humphries J, Borg-Bartolo B, Haddock M. Comparative proteomic analysis of supportive and unsupportive extracellular matrix substrates for human embryonic stem cell maintenance. J Biol Chem. 2013;26:18716–31.

38. Rebekah L, Gundry W, Kenneth R. Pluripotent stem cell heterogeneity and the evolving role of proteomic technologies in stem cell biology. Proteomics. 2011;11:3947–61.
39. Ong SE, Mann M. Mass spectrometry-based proteomics turns quantitative. Nat Chem Biol. 2005;1:252–62.
40. Ong SE, Blagoev B, Kratchmarova I, Kristensen DB, Steen H, Pandey A, Mann M. Stable isotope labeling by amino acids in cell culture, SILAC, as a simple and accurate approach to expression proteomics. Mol Cell Proteomics. 2002;1:376–86.
41. Park HW, Shin JS, Kim CW. Proteome of mesenchymal stem cells. Proteomics. 2007;7:2881–94.
42. Pinkse MWH, Uitto PM, Hilhorst MJ, Ooms B, Heck AJR. Selective isolation at the femtomole level of phosphopeptides from proteolytic digests using 2D-Nano LC-ESI-MS/MS and titanium oxide precolumns. Anal Chem. 2004;76:3935–43.
43. Pinkse MWH, Mohammed S, Gouw JW, Van Breukelen B, Vos HR, Heck AJR. Highly robust, automated, and sensitive online TiO2-based phosphoproteomics applied to study endogenous phosphorylation in Drosophila melanogaster. J Proteome Res. 2008;7:687–97.
44. Prowse AB, McQuade LR, Bryant KJ, Van Dyk DD, Tuch BE, Gray PP. Proteome analysis of conditioned media from human neonatal fibroblasts used in the maintenance of human embryonic stem cells. Proteomics. 2005;5:978–89.
45. Scott IC, Clark TG, Takahara K, Hoffman GG, Greenspan DS. Structural organization and expression patterns of the human and mouse genes for the type I procollagen COOH-terminal proteinase enhancer protein. Genomics. 1999;55:229–34.
46. Shevinsky LH, Knowles BB, Damjanov I, Solter D. Monoclonal antibody to murine embryos defines a stage-specific embryonic antigen expressed on mouse embryos and human teratocarcinoma cells. Cell. 1982;30:697–705.
47. Speers AE, Wu CC. Proteomics of integral membrane proteins theory and application. Chem Rev. 2007;107:3687–714.
48. Spisak S, Tulassay Z, Molnar B, Guttman A. Protein microchips in biomedicine and biomarker discovery. Electrophoresis. 2007;28:4261–73.
49. Stanton LW, Bakre MM. Genomic and proteomic characterization of embryonic stem cells. Curr Opin Chem Biol. 2007;11:399–404.
50. Sze SK, de Kleijn DP, Lai RC, Khia Way TE, Zhao H, Yeo KS, Low TY, Lian Q, Lee CN, Mitchell W, El Oakley RM, Lim SK. Elucidating the secretion proteome of human ESC derived mesenchymal stem cells. Mol Cell Proteomics. 2007;6:1680–9.
51. Van Hoof D, Passier R, Ward-Van Oostwaard D, Pinkse MWH, Heck AJR, Mummery CL, Krijgsveld J. A quest for human and mouse embryonic stem cell-specific proteins. Mol Cell Proteomics. 2006;5:1261–73.
52. Van Hoof D, Mummery CL, Heck AJR, Krijgsveld J. Embryonic stem cell proteomics. Proteomics. 2006;3:427–37.
53. Van Hoof D, Pinkse MWH, Ward-Van Oostwaard D, Mummery CL, Heck AJR, Krijgsveld J. An experimental correction for arginine-to-proline conversion artifacts in SILACbased quantitative proteomics. Nat Methods. 2007;4:677–8.
54. Washburn MP, Wolters D, Yates JR. Large-scale analysis of the yeast proteome by multidimensional protein identification technology. Nat Biotechnol. 2001;19:242–7.
55. Wegner M, Stolt CC. From stem cells to neurons and glia: a Soxist's view of neural development. Trends Neurosci. 2005;28:583–8.
56. Wu Q, Chen X, Zhang J, Loh YH, Low TY, Zhang W, Zhang W, Sze SK, Lim B, Ng HH. Sall4 interacts with Nanog and co-occupies Nanog genomic sites in embryonic stem cells. J Biol Chem. 2006;281:24090–4.
57. Xie CQ, Lin G, Luo KL, Luo SW, Lu GX. Newly expressed proteins of mouse embryonic fibroblasts irradiated to be inactive. Biochem Biophys Res Commun. 2004;315:581–8.
58. Zhang J, Tam WL, Tong GQ, Wu Q, Chan HY, Soh BS, Lou Y, Yang J, Ma Y, Chai L, Ng HH, Lufkin T, Robson P, Lim B. Sall4 modulates embryonic stem cell pluripotency and early embryonic development by the transcriptional regulation of Pou5f1. Nat Cell Biol. 2006;8:1114–23.

59. Zhao S, Nichols J, Smith AG, Li M. SoxB transcription factors specify neuroectodermal lineage choice in ES cells. Mol Cell Neurosci. 2004;27:332–42.
60. Mitsui K, Tokuzawa Y, Itoh H, Segawa K. The homeoprotein Nanog is required for maintenance of pluripotency in mouse epiblast and ES cells. Cell. 2003;113:631–42.
61. Domon B, Aebersold R. Mass spectrometry and protein analysis. Science. 2006;312:212–7.
62. Kelleher NL. Top-down proteomics. Anal Chem. 2004;76:196–203.
63. Siuti N, Kelleher NL. Decoding protein modifications using top-down mass spectrometry. Nat Methods. 2007;4:817–21.
64. Forner F, Foster LJ, Toppo S. Mass spectrometry data analysis in the proteomics era. Curr Bioinforma. 2007;2:63–93.
65. Lam H, Deutsch EW, Eddes JS, Eng JK, et al. Building consensus spectral libraries for peptide identification in proteomics. Nat Methods. 2008;5:873–5.
66. Bendall SC, Hughes C, Campbell JL, Stewart MH, et al. An enhanced mass spectrometry approach reveals human embryonic stem cell growth factors in culture. Mol Cell Proteomics. 2008;8:421–32.
67. Fang Y, Robinson DP, Foster LJ. Quantitative analysis of proteome coverage and recovery rates for upstream fractionation methods in proteomics. J Proteome Res. 2010;9:1902–12.
68. Mueller LN, Brusniak MY, Mani DR, Aebersold R. An assessment of software solutions for the analysis of mass spectrometry based quantitative proteomics data. J Proteome Res. 2008;7:51–61.
69. Domon B, Aebersold R. Options and considerations when selecting a quantitative proteomics strategy. Nat Biotechnol. 2010;28:710–21.
70. Mann M. Functional and quantitative proteomics using SILAC. Nat Rev Mol Cell Biol. 2006;7:952–8.
71. Ross PL, Huang YLN, Marchese JN, Williamson B, et al. Multiplexed protein quantitation in Saccharomyces cerevisiae using amine-reactive isobaric tagging reagents. Mol Cell Proteomics. 2004;3:1154–69.
72. Bryder D, Rossi DJ, Weissman IL. Hematopoietic stem cells: the paradigmatic tissue-specific stem cell. Am J Pathol. 2006;169:338–46.
73. Kern S, Eichler H, Stove J, Kluter H, Bieback K. Comparative analysis of mesenchymal stem cells from bone marrow, umbilical cord blood, or adipose tissue. Stem Cells. 2006;24:1294–301.
74. Pevsner-Fischer M, Levin S, Zipori D. The origins of mesenchymal stromal cell heterogeneity. Stem Cell Rev. 2011;7:560–8.
75. Davies OG, Cooper PR, Shelton RM, Smith AJ, Scheven BA. A comparison of the in vitro mineralization and dentinogenic potential of mesenchymal stem cells derived from adipose tissue, bone marrow, and dental pulp. J Bone Miner Metab. 2014;33:371–82.
76. Boyd NL, Robbins KR, Dhara SK, West FD, Stice SL. Human embryonic stem cell-derived mesoderm-like epithelium transitions to mesenchymal progenitor cells. Tissue Eng. 2009;15:1897–907.
77. Hematti P. Human embryonic stem cell-derived mesenchymal stromal cells. Transfusion. 2011;51:138S–44S.
78. de Peppo GM, et al. Human embryonic mesodermal progenitors highly resemble human mesenchymal stem cells and display high potential for tissue engineering applications. Tissue Eng. 2010;16:2161–82.
79. Vodyanik MA, et al. A mesoderm-derived precursor for mesenchymal stem and endothelial cells. Cell Stem Cell. 2010;7:718–29.
80. Kimbrel EA, et al. Mesenchymal stem cell population derived from human pluripotent stem cells displays potent immunomodulatory and therapeutic properties. Stem Cells Dev. 2014;23:1611–24.
81. Sanchez L, et al. Enrichment of human ESC-derived multipotent mesenchymal stem cells with immunosuppressive and anti-inflammatory properties capable to protect against experimental inflammatory bowel disease. Stem Cells. 2011;29:251–62.

82. Wang X, et al. Human ESC-derived MSCs outperform bone marrow MSCs in the treatment of an EAE model of multiple sclerosis. Stem Cell Rep. 2014;3:115–30.
83. Liberski AR, et al. Adaptation of a commonly used, chemically defined medium for human embryonic stem cells to stable isotope labeling with amino acids in cell culture. J Proteome Res. 2013;12:3233–45.
84. Yamana R, Iwasaki M, Wakabayashi M, Nakagawa M, Yamanaka S, Ishihama Y. Rapid and deep profiling of human induced pluripotent stem cell proteome by one-shot NanoLC–MS/MS analysis with meter-scale monolithic silica columns. J Proteome Res. 2012;44:1–12.
85. Chae J, Kim D, Lee N, Jeon Y, Jeon I, Kwon J, Kim J, Soh Y, Lee D, Seo K, Choi N, Park B, Kang S, Ryu J, Oh S, Shin D, Lee D, Do J, Park I, Daley G, Song J. Quantitative proteomic analysis of induced pluripotent stem cells derived from a human Huntington's disease patient. Biochemist. 2012;446:359–71.
86. Pripuzova N, Getie-Kebtie M, Grunseich C, Sweeneyc C, Malechc H, Alterman M. Development of a protein marker panel for characterization of human induced pluripotent stem cells (hiPSCs) using global quantitative proteome analysis. Stem Cell Res. 2015;14:323–38.

Lipidomics of Adipogenic Differentiation of Mesenchymal Stem Cells

Kambiz Gilany, Moloud Payab, Parisa Goodarzi, Akram Tayanloo-Beik, Sepideh Alavi-Moghadam, Maryamossadat Mousavi, Babak Arjmand, Tannaz Safaralizadeh, Mina Abedi, Maryam Arabi, Hamid Reza Aghayan, and Bagher Larijani

Abstract Mesenchymal stem cells are defined as multipotent cells which have the ability to differentiate into various types of cell. Under the adipogenic stimuli, mesenchymal stem cells possess the ability to differentiate into adipocytes through adipogenesis processes. Adipogenesis is defined as the process of pre-adipocyte differentiation to mature adipocytes. Adipocytes are a type of cells with the ability to maintain energy balance through storage excess energy. However, several abnormal conditions including various types of disease can result from energy imbalance. Accordingly, obesity as a worldwide problem is one of the prevalent outcomes of

---

K. Gilany
Department of Biomedical Sciences, University of Antwerp, Antwerp, Belgium

Integrative Oncology Department, Breast Cancer Research Center,
Motamed Cancer Institute, ACECR, Tehran, Iran

M. Payab
Obesity and Eating Habits Research Center, Endocrinology and Metabolism
Molecular-Cellular Sciences Institute, Tehran University of Medical Sciences, Tehran, Iran

P. Goodarzi
Brain and Spinal Cord Injury Research Center, Neuroscience Institute,
Tehran University of Medical Sciences, Tehran, Iran

A. Tayanloo-Beik · S. Alavi-Moghadam · T. Safaralizadeh · H. R. Aghayan
Cell Therapy and Regenerative Medicine Research Center, Endocrinology and Metabolism
Molecular-Cellular Sciences Institute, Tehran University of Medical Sciences, Tehran, Iran

M. Mousavi · M. Abedi · M. Arabi
Metabolomics and Genomics Research Center, Endocrinology and Metabolism
Molecular-Cellular Sciences Institute, Tehran University of Medical Sciences, Tehran, Iran

B. Arjmand
Cell Therapy and Regenerative Medicine Research Center, Endocrinology and Metabolism
Molecular-Cellular Sciences Institute, Tehran University of Medical Sciences, Tehran, Iran

Metabolomics and Genomics Research Center, Endocrinology and Metabolism
Molecular-Cellular Sciences Institute, Tehran University of Medical Sciences, Tehran, Iran

B. Larijani (✉)
Endocrinology and Metabolism Research Center, Endocrinology and Metabolism
Clinical Sciences Institute, Tehran University of Medical Sciences, Tehran, Iran

B. Arjmand (ed.), *Genomics, Proteomics, and Metabolomics*, Stem Cell Biology and Regenerative Medicine, https://doi.org/10.1007/978-3-030-27727-7_7

increasing fat mass and there is a global effort to combat it. Accumulation of excess fat in adipocytes leads to adipocyte hypertrophy. Consequently, hypertrophic adipocyte can secrete several endocrine factors that induce hyperplasia (one of the major causes of obesity) and signal proliferation and differentiation of pre-adipocytes. Therefore, it has been demonstrated that adipogenic differentiation of mesenchymal stem cells undergoes different signaling pathways with various regulatory factors, while elucidation of these controllers can help scientists to develop more effective treatments for obesity and other related diseases. Therein, lipids have been presented as pivotal mediators of cellular processes and could induce several signaling pathways. Additionally, lipids are fundamental metabolites which use as cellular biomarkers to indicate different biological states and cellular activity. Total content of lipids in cells is known as lipidome. Any slight changes in the lipidome reflect different cellular changes. Tracking and comparing these changes between different stages of mesenchymal stem cell differentiation can provide identification of essential metabolic pathways involved in adipogenesis. In this context, lipidomics has been introduced as an emerging field of stem cell and regenerative medicine. Through the large-scale analysis of lipids, lipidomics provides more efficient methods to the investigation of adipocytes, and also prediction of the prognosis of obesity and its prevention and treatment.

Keywords Biomarkers · Lipids · Lipidomics · Mesenchymal stem cell · Obesity

Abbreviations

BAT	Brown adipose tissue
BMI	Body mass index
C/EBPα	CCAAT/enhancer binding proteins
C1	Carbon 1
C2	Carbon 2
C3	Carbon 3
CKI	Cyclin-dependent kinase inhibitors
CoA	Coenzyme A
Dex	Dexamethasone
FA	Fatty acids
GD	Growth arrest
GL	Glycerolipids
hMSCs	Human mesenchymal stem cells
IBMX	Isobutylmethylxanthine
IBMX	Isobutyl-methylxanthine
IL6	Interleukin 6
ISCT	International Society for Cell Therapy
LD	Lipid droplets
MCE	Clonal expansion

MSCs	Mesenchymal stem cells
MVA	Mevalonic acid
PG	Prostaglandins
PPARγ	Peroxisome proliferator-activated receptor γ
PPARϒ	Peroxisome proliferator-activated receptor ϒ
PSCs	Pluripotent stem cells
SREBP	Sterol regulatory element binding protein
TAG	Triacylglycerol
TG	Triglycerides
TGF-β3	Transforming growth factor-β3
TNFα	Tumor necrosis factor α
WAT	White adipose tissue

1 Introduction

Various species of lipids (which are soluble in nonpolar solvents) as a class of the major biological molecules and metabolites have key function in cell energy storage, structure, and signaling [1, 2]. Accordingly, they can determine the cell fate through their bioactive roles including mediating inflammation, regulating the polarity of cellular membranes, inducing apoptosis, and maintaining the cell survival [3–6]. Indeed, lipid metabolites play their bioactive roles by several mechanisms such as direct interaction with specific protein binding partners, organization of lipid micro domains or rafts, and invocation of lipid second messengers to regulate the energy and redox balance for stem cell differentiation, proliferation, and hemostasis [7, 8]. In other words, one of the invaluable methods for obtaining information about the molecular level of the cell function is analyzing and monitoring the lipid metabolites [9–11]. Lipidomics as a new developed discipline is referred to a broad-scale study of lipid metabolism which can clarify the biochemical mechanisms underlying particular alterations in lipid metabolism [12, 13]. Lipid metabolism which can be regulated by various signaling pathways is a complex process which involves uptaking, transporting, synthesizing, and degrading. Moreover, it was demonstrated that the mechanisms of regulation can be different in similar lipids (depend on different cell and tissue conditions) [1, 14, 15]. On the other hand, the variation of lipid can lead to the diversity in gene expression, protein distribution and phenotyping, and membrane compositions in cells [16, 17]. For instance, based on lipid metabolism, mesenchymal stem cells (MSCs) (a group of multipotent stem cells with less ethical concerns and immunomodulatory features) are being used as therapeutic agents with great success for different diseases [18, 19]. Furthermore, in various differentiation conditions they have different lipidomic profiles. Accordingly, lipid metabolites monitoring of differentiated cells in comparison with undifferentiated ones can be useful to perform controlled differentiation for therapeutic purposes [18]. Hereupon, the objective of this chapter is to review the various features of the lipidomics of adipogenic differentiation in mesenchymal stem cells.

2 Mesenchymal Stem Cells

MSCs as multipotent stromal cells have two intrinsic properties similar to other stem cells, including differentiation and self-renewal ability (Fig. 1). Accordingly, they can be divided asymmetrically (by means of both external and internal factors) in which stem cell can produce a stem cell and a differentiated cell. On the other hand, they can be proliferated through symmetric division in which stem cell can produce only two stem cells or two differentiated cells [20–24] (Fig. 2). Moreover,

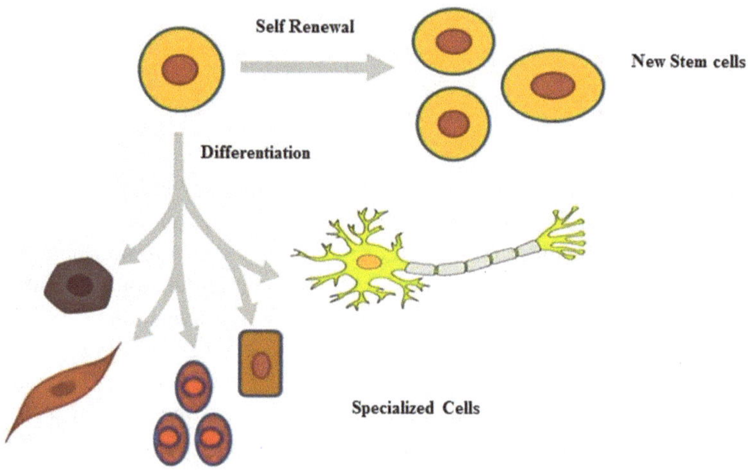

Fig. 1 Intrinsic properties of stem cells. Two intrinsic properties of stem cells are differentiation and self-renewal ability. Stem cells are uncommitted cells that could replenish their own undifferentiated cell numbers. When undergoing different stimuli, stem cells could differentiate into specialized cell types

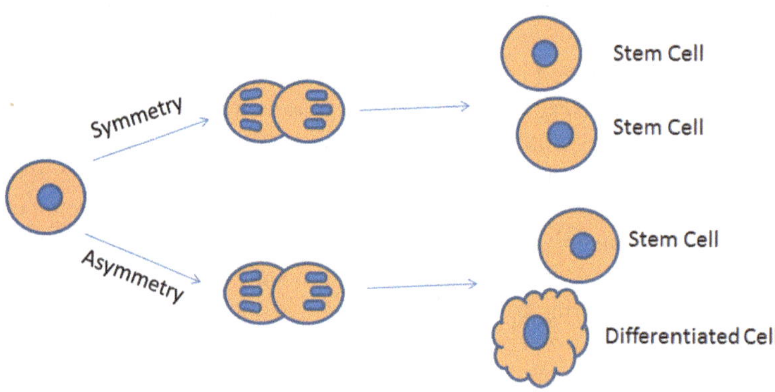

Fig. 2 Symmetric and asymmetric stem cell division

MSCs are known as valuable sources for regenerative medicine because of their specific properties. Hereupon, understanding the characteristics of MSCs can be useful in their cell-based treatment applications [25, 26].

3 Mesenchymal Stem Cell Characterization

MSCs were identified for the first time in 1966 by Fridenshtein et al. from bone marrow tissue. Today, it is known that MSCs can be isolated from various tissues containing the bone marrow, skin, umbilical cord, adipose tissue, menstrual blood, etc. [27–29]. In accordance with the minimal criteria which were developed by International Society for Cellular Therapy (ISCT), MSCs were defined by three characteristics, including showing plastic-adherent when using tissue culture dishes, expressing some special cell surface markers such as CD105, CD73, and CD90, and potential of being in vitro differentiate into multiple mesenchymal lineages such as osteoblasts, adipocytes, and chondrocytes [30–33].

4 Mesenchymal Stem Cell Differentiation

The differentiation potential of MSCs is associated with numerous agents containing mechanical signals and the physical responses of the cells [26, 34]. Therein, MSCs can be changed and exhibited relevant biomolecules in response to different environmental stimuli such as adipogenesis induction [35, 36]. In this context, for adipogenic induction of MSCs, combination of insulin, isobutylmethylxanthine (IBMX), dexamethasone (Dex), and a peroxisome proliferator-activated receptor γ (PPARγ) agonist is generally used. Subsequently, after 7 days of induction, for the evaluation of adipogenic phenotype, lipid droplets (LD) that accumulated in the cells can be stained with lipophilic dyes (e.g., oil red o) [37, 38]. Additionally, osteogenic differentiation of MSC can be induced by application of Dex, L-ascorbic acid, and β-glycerophosphate. For the evaluation of osteogenic differentiation, calcium depositions can be observed through Alizarin red or Von Kossa staining [39, 40]. Moreover, for chondrogenic differentiation of MSCs, the pelleted micromass of cells is cultured in serum free media supplemented with transforming growth factor-β3 (TGF-β3) and after induction period for confirming the presence of proteoglycans in the cell pellets alcian blue staining can be used [41–43].

5 Adipogenesis

5.1 Overview of Adipogenesis

Adipogenesis is the process of pre-adipocyte differentiation to mature fat cells with the ability of fat storage named adipocytes (Fig. 3). Adipogenesis is initiated with differentiation signals that result in adipocytes generation. Adipocytes are the major cell types in adipose tissue generation which have pivotal role in energy homeostasis and body metabolism [44]. In energy imbalance conditions, adipocytes store excess energy as triacylglycerol (TAG) in order to maintain energy balance that results in an elevation of adipose tissue mass. Consequently, the development of obesity could be one of the common outcomes of increasing in the adipose tissue mass. Furthermore, adipocytes are considered as an endocrine organ that secretes several factors [45, 46]. Secreted lipid and protein factors could strongly affect tissue metabolism, insulin sensitivity, different immunological response, and several types of disease. Some of globally prevalent diseases like obesity and type 2 diabetes mellitus were drown special attention in recent decades [47, 48]. Especially, considering the strong link between obesity and metabolic disorders, there is an

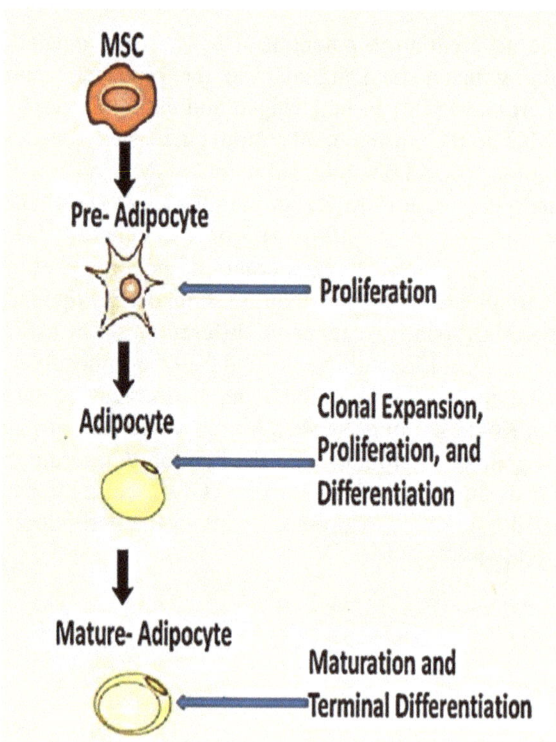

Fig. 3 Adipogenesis. The process of MSC differentiation in which MSCs are differentiated to pre-adipocytes and consequently to adipocytes through some mechanism including proliferation, clonal expansion, and maturation [44]

increasing need for identification of the key mechanisms involved in energy balance. Therefore, understanding the adipose cells development mechanisms possess a great importance to manage such diseases [46, 49].

5.2 *Cell Line Models for Adipocyte Differentiation*

Nowadays, specific cell line derived from mouse 3T3 cells is most commonly used for adipogenesis process investigation. 3T3-L1 cell lines are committed to adipocyte lineage and under the different adipogenic stimuli, they are capable to differentiate into adipocytes. Hence, it is not considering as the best choice for adipogenic studies, especially for the understanding of committing pathways. Furthermore, according to the body of literature, there are some differences in adipogenic developmental process of murine and human origin cell lines. For instance, different origins of 3T3-L1 express various types of Sterol Regulatory Element Binding Protein (SREBP). Therein, in adipogenesis process, ADD/SREBP-1c (expresses by human) is more involved than SREBP-1a (from murine origin) [50, 51]. Despite that PPARs are the more important regulators of adipogenesis compared to SREBPs, existing differences between various species in expression of SREBP types could affect our knowledge about human adipogenesis processes [52]. In this regard, human MSCs (hMSCs) have been introduced as a suitable substitute for investigating adipogenesis processes in human. hMSCs are multipotent cells that possess the potential of differentiating to adipocytes. In other words, they are uncommitted cells that are suitable for discovery of unknown genes and factors which are involved in commitment to adipogenesis processes [53, 54]. However, although hMSCs seem more appropriate model in this field which has been identified for more than two decades, they were not advantageous to figure out the adipogenesis biological mechanisms due to difficulty in the extraction of homogeneous MSCs. Therefore, 3 T3-L1 still possesses its specific impact on adipogenesis description [50, 55, 56]. Hereupon, some controlling mechanisms including commitment, growth arrest, mitotic clonal expansion, and terminal differentiation direct the adipocyte differentiation. Commitment is an essential process of stem cell differentiation to the specific lineages [57, 58]. Through the commitment process, stem cells receive different signals which determine the differentiation to specific lineages, adipogenic lineage, for instance. It has been suggested that mature adipocytes secret signal molecules like Wnt which can induce cells to undergo adipogenesis. Additionally, high expression of Wnt signaling pathway in pre-adipocytes and its reduction during differentiation demonstrate its importance in the commitment process. Though, more studies are needed to elucidate the adipocyte-specific factors involving in this process [59, 60]. Growth arrest as another controlling event has been accepted that occurs two times during developmental process of adipocytes, before clonal expansion (MCE) and prior to terminal adipocyte differentiation. In the first mitotic arrest, it has been demonstrated that cyclin-dependent kinase inhibitors (CKI), p21 and p27 undergo a notable increase. While, during the second growth arrest, p18 that is a type of CKI represents a significant raise. Additionally, a correlation between mitotic arrest and differentiation has been proposed by implying the role of PPARγ in regulating the expression of CKI. Continually, mitotic clonal expansion occurs once mitogenic and adipogenic signals are received by cells which result in multifold DNA replication. During DNA replication, chromatin structure undergoing some changes allows transcription factors to access binding sites. These processes altogether could allow a wide range of necessary genes to turn adipogenesis on or off. After clonal expansion, given cells undergo second growth arrest (GD) in which there is no point of return, where cells are committed to undergo adipogenesis [58, 61, 62].

6 Lipid Classifications and Functional Properties

Several types of lipids have been shown to induce cell signaling pathways and provide health profits which they considered as bioactive lipids. They are generated from the metabolism of cellular membrane elements. In other words, lipids are not only the main components of cell membranes and effective energy sources, but also fundamental mediators for different cellular processes [63, 64]. Biologically, lipid is a soluble molecule in nonpolar solvents. In a broad definition, lipids are known as hydrophobic or amphiphilic molecules. Biochemically, keto-acyl and isoprene are two distinct types of lipid subunits that form lipids. Accordingly, lipids can be divided into eight categories [65–67]: fatty acids (FA) (organic acids with the functional group–COOH which are categorized as one of the essential lipid groups and possess specific importance as building-block of lipids) [68, 69], glycerolipids (GL) (which are structured through FA and glycerol link by ester bonds and have a pivotal role in the membrane [70], glycerophospholipids (which are formed through the glycosidic attachment of some sugar residues to glycerol and they form lipid bilayers spontaneously which shape double layer membrane) [17, 71–73], sphingolipids (a distinct class of bioactive lipids with a backbone of sphingoid bases which are the important elements of the cell membrane and it is demonstrated that simple sphingolipid metabolites be potential mediators in cell inflammation, autophagy, necrosis, apoptosis, senescence, stress, proliferation, and differentiation signaling cascades) [74–76], saccharolipids (the classes of lipids in which FA are joined to sugar parts and they have a compatible structure with membrane bilayers) [77, 78], polyketides (a broad group of secondary metabolites which are organized through the condensation of acetyl-coenzyme A (CoA) and malonyl-CoA with antimicrobial and immunosuppressive features) [79, 80], sterol lipids alcohols (which are derived from the fused four-ring core structure and as are organic molecules which can control biological processes and maintain the structure of eukaryotic cells) [76, 81], and prenol lipids (the type of lipids which are synthesized from the five-carbon-unit precursors isopentenyl diphosphate and dimethylallyl diphosphate derived from mevalonate through the mevalonic acid (MVA) pathway) [66, 82, 83]. Furthermore, lipids are used as cellular biomarkers to indicate the biological state and clinical diagnostic tools. Therein, monitoring and quantification of the whole lipids of cells (lipidome) can provide insights into the particular lipid activities in cells and also to set up therapeutic or preventive programs for various diseases [84, 85].

7 Relationship between Lipids and Adipogenesis

The use of 3T3-L1 provides useful information on lipid profiling and biological pathways. This cell line first has a fibroblast-like morphology, and then, under certain conditions, it can be differentiated into adipocyte phenotype. Considering 3T3-L1, it is possible to identify effective lipid groups in the development of adipocytes in MSCs [86]. Investigating the results leads to the identification of effective pathways for adipocyte development and consequently, the progression of adipogenesis. FA, depending on the length of the carbon chain, the degree of unsaturation, and the location of the double bond, have an incremental or decreasing effect on the adipogenesis. For example, medium chain FA such as decanoic acid and octanoic acid lead to an increase in lipid levels in 3T3-L1 cells [87]. It has been demonstrated that a category of polyunsaturated omega 3 such as docosahexaenoic acid reduces the adipogenesis by reducing the expression of the adipogenic gene in 3T3-L1 cells [88]. Linoleic acid leads to an increase in the formation of lipid in the cells. The other group of lipids is lysophosphatidic acid that are bound to the LPA1 receptor and can decrease expression and activity of the PPARγ, and ultimately prevents the expression of the adipogenic genes. The next group of lipids are sterols, which have an association with adipogenesis. Among this category, oxy sterols prevent adipogenesis and studies have been done to investigate their effect, which suggests the recognition of oxysterols as an effective and important factor in reducing adipogenesis and increasing osteogenesis [89, 90]. Finally, here mentioned just some lipids that are effective in the adipogenesis process, which can be very helpful in treating and reducing obesity.

8 Obesity and Related Diseases

Obesity is the condition of accumulation of excess body fat which has adverse effects on human health. Globally, it is considered as the main problem of the health care system. Obesity is a worldwide epidemic obstacle and rising not only in developed countries but also it is a well-known problem in developing societies. Nowadays, body mass index (BMI) is used as the gold standard to determine the obesity level according to defined ranges ($\geq$30 kg/m^2 for obese and 25–30 kg/m^2 for overweight). However, the BMI value is variable in some countries, for instance, it is lower in some East Asian countries [91, 92]. Obesity could cause an increase in the risk of several diseases like cardiovascular diseases, obstructive sleep apnea, osteoarthritis, type 2 diabetes, and cancer [93–95]. Furthermore, obesity is one of the leading causes of death worldwide but fortunately preventable [96, 97]. According to epidemiological studies, obesity comprises a wide range of individuals from children to adults, and it is more prevalent in women [98–100]. Related authorities point it as one of the major burdens of public health in twenty-first century [101]. Historically, obesity was a sign of health, wealth, and fertility, even though nowadays it is categorized as a disease and in the most of civilizations it has been stigmatized. Typically, changing of dietary habits and promoting of the sedentary lifestyle result in rising TAG storage in adipose tissue due to excess accumulation of energy and energy imbalance. Excess energy storage in adipose tissue causes adipocyte hypertrophy (adipocyte enlargement). Furthermore, an increase in the number of adipocytes results in hyperplasia that is considered as one of the main causes of extreme form of obesity. Hyperplasia is referred to as the pre-adipocytes differentiation to mature adipocytes in the adipogenesis process. It has been shown that mature adipocytes as an endocrine organ can secret paracrine factors that lead to hyperplasia. Moreover, paracrine factors secreted by mature adipocytes could influence pre-adipocytes differentiation [102–104].

8.1 The Role of Mesenchymal Stem Cells in Obesity

Obesity occurs with extensive adipose tissue which can produce two mechanisms: adipocyte hypertrophy and adipocyte hyperplasia [105]. These procedures are causing enrichment of size and an increase in the number of fat cells, respectively [106]. Increasing the fat mass can stimulate adipocytes hypertrophy, as the main element in reserving lipids and then be continued by hyperplasia [107, 108]. Hypertrophic adipocytes act like an endocrine organ [109, 110]. Accordingly, this organ can secret various factors that have substantial roles in proliferation and differentiation of progenitor cells to adipocyte [111]. These factors include complement factors, peptides, and cytokines that act to regulate adipocyte metabolism and growth. In addition, adipokine proteins are secreted by mature adipose tissue that include resistin, leptin, adiponectin, interleukin 6 (IL6), and tumor necrosis factor α (TNFα), which can promote hyperplasia [112]. These mechanisms are influenced by genetic and diet that suggests many factors can be activated and committed to obesity [113]. Generally, adipogenesis process contains two steps. At first, pluripotent stem cells (PSCs) change to MSCs which can be differetiated into pre-adipocytes and then pre-adipocytes be differentiated into mature adipocytes [114]. Pre-adipocyte is differentiated into two types of cells, white adipocyte which can make white adipose tissue (WAT) (stored triglycerides (TG)) and brown adipocyte which can produce brown adipose tissue [87] (burning fat and produce heat) [115, 116]. Indeed, WAT is principal in fat storage as a TG which have important role in obesity. Based on some reports, the progenitor of white adipocytes has been detected in the adipose vasculature (perivascular cells). Accordingly, adipose vasculature can be assumed as niche of progenitors and produces proper signal for the development of adipocytes [117, 118]. Different factors influence on the MSC adipogenic development. Two main transcription factors are the peroxisome proliferator-activated receptor ϒ (PPARϒ) and the CCAAT/enhancer binding proteins (C/EBPα) that lead adipocyte differentiation to white adipocytes in the early step. Herein, it has been considered that PPARϒ is the main factor in the regulation of adipogenesis processes. In other words, PPARϒ plays a crucial role in the pre-adipocyte differentiation to adipocyte and finally mature adipocytes. In this regards, PPARϒ can be investigated as a key target for the treatment of obesity and other related diseases [119]. Eventually, understanding the mechanisms underlying MSCs differentiation to pre-adipocytes and the development of pre-adipocytes to mature adipocytes are comprised of several regulators such as multiple transcription factors that need to be elucidated. For this purpose, MSCs could be a promising option for control of different adipogenesis pathways and clarify the relation of fundamental MSCs adipogenesis mechanisms and obesity. Generally, it is anticipated that MSC-based studies could ultimately contribute to treating obesity and other metabolic diseases in the future [114, 120, 121].

9 Lipidomics

Lipidomics is one of the branches of OMICs, as an emerging technology in the biological sciences. This branch of OMICs has been raised after the genomics and proteomics, which is considered as a subgroup of metabolomics [122, 123]. This technology examines lipids in large-scale based on chemical and analytical methods, in particular mass spectrometry and emerged in 2003. Some of the methods that allow large-scale study of lipids have accelerated the progression of lipidomics, like mass spectrometry. Mass spectrometry is highly applicable due to its high analytical power and rapid growth of its technology. Lipidomics is a missing part in the study of MSCs that is of great importance in understanding the MSCs biology. Lipids are essential metabolites that carry out a lot of cellular activity. Lipidome is the total lipid content of a cell. Any slight changes in the lipidome content can simultaneously reflect changes in the enzymes level and the gene expression pattern [124, 125]. Comparing lipid profile changes between undifferentiated and differentiated MSCs can provide recognition of adipogenic effective lipids. Tracking these changes leads to the identification of metabolic pathways associated with these lipids [126]. By examining metabolic pathways that affect adipogenesis, interventional strategies to prevent obesity can be identified. Ultimately, getting a lipid profile will make the adipocyte study much smoother and provide effective ways to reduce obesity (Fig. 4) [128, 129].

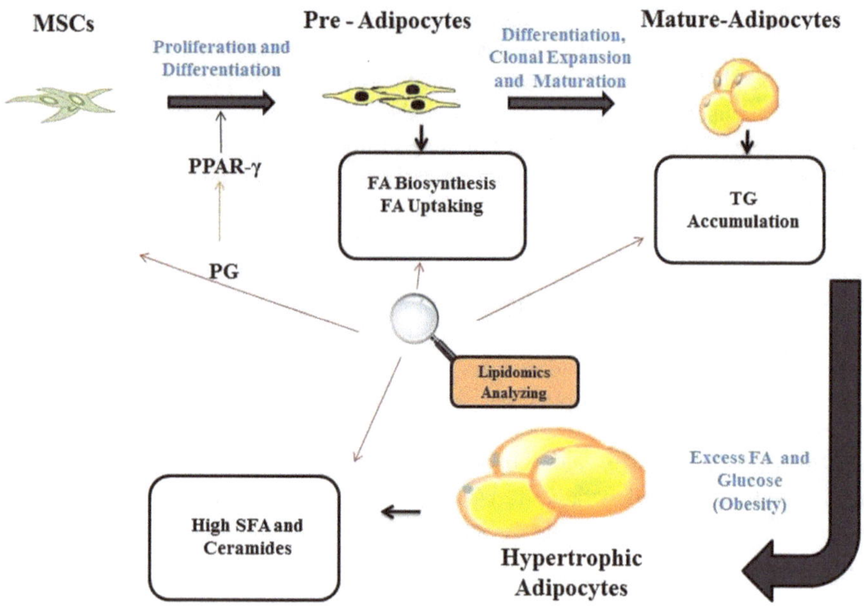

Fig. 4 Lipidomics analyzing and adipogenesis. PPARγ which is required for adipogenesis and regulated by prostaglandins (PGs) can induce expression of genes involved in FA uptake and lipid storage. MSCs are differentiated into adipocytes through the adipogenesis. Lipids (especially TG) are accumulated in (pre)adipocytes by fatty acid uptaking and biosynthesis. On the other hand, excessive lipid and glucose accumulation (obesity state) can lead to adipocyte hypertrophy (with high expression of saturated FA and ceramides). Lipidomics as a functional and diagnostic tool can distinguish and compare the lipidomic profile of different cells [127]

10 Conclusion and Future Perspective

Recently, lipidomics analyzing and unbiased global probing of lipid alternations in different cells such as MSCs have become a crucial field of metabolomics to implement in various clinical areas [130–132]. In this context, the evaluation of MSCs differentiation into adipocytes represents an impressive method for the study on their post-differentiation immunological characteristics [19, 133, 134]. Accordingly, differentiated cells can secrete hormones (which are known to show potent immunomodulatory impacts) and also provide pro-inflammatory cytokines and chemokines [134]. Hereupon, due to some reports, the outcome of adipogenesis on immunoregulatory activity seems to be tissue-specific, based on different tissue abilities. For instance, according to monitoring of lipidomics, through the adipogenesis, non-adipose tissue derived MSCs miss some of their immunoprotective impressions [127, 135]. Generally, the lipidomic profile of undifferentiated cells is different compared with the differentiated cells in each lineage stages [127, 136]. In other words, the lipidomic profile of adipocytes differentiated from pre-adipocytes is considerably different from the adipocytes differentiated from mesenchymal stem cells. Indeed, multiple in vivo signals driving lipogenesis and various in vitro conditions can lead to these differences in lipidomic profiles [105, 137, 138]. Besides that, lipidomics can distinguish between different forms of adipose tissues (WAT and BAT) in accordance with their particular lipidome. Collectively, lipidomics as a precious tool can decipher lipidomic profile in a specific experimental and therapeutic settings and recognize bio-active lipids and lipid metabolism alternations. Additionally, due to the tight association of lipids with metabolic disorders including obesity, lipidomics has a crucial role in medical monitoring and mechanistic investigations for metabolic diseases [124]. On the other hand, it seems that in the near future lipidomics in combination with other OMICs approaches such as genomics and proteomics can be used for identifying particular cell types, as well as applying in the field of personalized medicine [86, 127, 139, 140].

References

1. Huang C, Freter C. Lipid metabolism, apoptosis and cancer therapy. Int J Mol Sci. 2015;16(1):924–49.
2. Rolim AE, et al. Lipidomics in the study of lipid metabolism: current perspectives in the omic sciences. Gene. 2015;554:131–9.
3. Kiamehr M, et al. Lipidomic profiling of patient-specific iPSC-derived hepatocyte-like cells. Dis Model Mech. 2017;10(9):1141–53.
4. Pébay A, Wong RC. Lipidomics of stem cells. New York: Springer; 2017.
5. Bieberich E, Wang G. Bioactive lipids in stem cell differentiation, in embryonic stem cells-differentiation and pluripotent alternatives. London: IntechOpen; 2011.
6. van Meer G, Voelker DR, Feigenson GW. Membrane lipids: where they are and how they behave. Nat Rev Mol Cell Biol. 2008;9(2):112–24.
7. Bieberich E. It's a lipid's world: bioactive lipid metabolism and signaling in neural stem cell differentiation. Neurochem Res. 2012;37(6):1208–29.
8. de Meyer FJM, et al. Molecular simulation of the effect of cholesterol on lipid-mediated protein-protein interactions. Biophys J. 2010;99(11):3629–38.
9. van Meer G. Cellular lipidomics. EMBO J. 2005;24(18):3159–65.
10. Harkewicz R, Dennis EA. Applications of mass spectrometry to lipids and membranes. Annu Rev Biochem. 2011;80:301–25.
11. Dowhan W, Mileykovskaya E, Bogdanov M. Diversity and versatility of lipid-protein interactions revealed by molecular genetic approaches. Biochim Biophys Acta. 2004;1666(1–2):19–39.
12. Wenk MR. The emerging field of lipidomics. Nat Rev Drug Discov. 2005;4(7):594.
13. Shevchenko A, Simons K. Lipidomics: coming to grips with lipid diversity. Nat Rev Mol Cell Biol. 2010;11(8):593.
14. Vance JE, Vance DE. Biochemistry of lipids, lipoproteins and membranes. Amsterdam: Elsevier; 2008.
15. Ridgway N, McLeod R. Biochemistry of lipids, lipoproteins and membranes. Amsterdam: Elsevier; 2015.
16. Dowhan W, Mileykovskaya E, Bogdanov M. Diversity and versatility of lipid–protein interactions revealed by molecular genetic approaches. Biochim Biophys Acta Biomembr. 2004;1666(1):19–39.
17. Harayama T, Riezman H. Understanding the diversity of membrane lipid composition. Nat Rev Mol Cell Biol. 2018;19:281.

18. Chatgilialoglu A, et al. Restored in vivo-like membrane lipidomics positively influence in vitro features of cultured mesenchymal stromal/stem cells derived from human placenta. Stem Cell Res Ther. 2017;8(1):31.
19. Campos AM, et al. Lipidomics of mesenchymal stromal cells: understanding the adaptation of phospholipid profile in response to pro-inflammatory cytokines. J Cell Physiol. 2016;231(5):1024–32.
20. Goodarzi P, et al. Therapeutic abortion and ectopic pregnancy: alternative sources for fetal stem cell research and therapy in Iran as an Islamic country. Cell Tissue Bank. 2019;20(1):11–24.
21. Shirian S, et al. Comparison of capability of human bone marrow mesenchymal stem cells and endometrial stem cells to differentiate into motor neurons on electrospun poly (ε-caprolactone) scaffold. Mol Neurobiol. 2016;53(8):5278–87.
22. Larijani B, et al. Human fetal skin fibroblasts: extremely potent and allogenic candidates for treatment of diabetic wounds. Med Hypotheses. 2015;84(6):577–9.
23. Goodarzi P, et al. Stem cell-based approach for the treatment of Parkinson's disease. Med J Islam Repub Iran. 2015;29:168.
24. Goodarzi P, et al. Stem cell therapy for treatment of epilepsy. Acta Med Iran. 2014;52(9):651–5.
25. Wang M, Yuan Q, Xie L. Mesenchymal stem cell-based immunomodulation: properties and clinical application. Stem Cells Int. 2018;2018:3057624.
26. Rohban R, Pieber TR. Mesenchymal stem and progenitor cells in regeneration: tissue specificity and regenerative potential. Stem Cells Int. 2017;2017:5173732.
27. Larijani B, et al. GMP-grade human fetal liver-derived mesenchymal stem cells for clinical transplantation. In: Stem cells and good manufacturing practices. New York: Springer; 2014. p. 123–36.
28. Derakhshanrad N, et al. Case report: combination therapy with mesenchymal stem cells and granulocyte-colony stimulating factor in a case of spinal cord injury. Basic Clin Neurosci. 2015;6(4):299.
29. Goodarzi P, et al. Adipose tissue-derived stromal cells for wound healing. Adv Exp Med Biol. 2018;1119:133–49.
30. Minguell JJ, Erices A, Conget P. Mesenchymal stem cells. Exp Biol Med. 2001;226(6):507–20.
31. Kassem M. Mesenchymal stem cells: biological characteristics and potential clinical applications. Cloning Stem Cells. 2004;6(4):369–74.
32. Mahmood R, Shaukat M, Choudhery MS. Biological properties of mesenchymal stem cells derived from adipose tissue, umbilical cord tissue and bone marrow. AIMS Cell Tissue Eng. 2018;2(2):78–90.
33. Ma J, et al. Comparative analysis of mesenchymal stem cells derived from amniotic membrane, umbilical cord, and chorionic plate under serum-free condition. Stem Cell Res Ther. 2019;10(1):19.
34. Sheykhhasan M, et al. Mesenchymal stem cells as a valuable agent in osteoarthritis treatment. Stem Cell Investig. 2018;5:41.
35. Phelps J, et al. Bioprocessing of mesenchymal stem cells and their derivatives: toward cell-free therapeutics. Stem Cells Int. 2018;2018:9415367.
36. Anderson HJ, et al. Mesenchymal stem cell fate: applying biomaterials for control of stem cell behavior. Front Bioeng Biotechnol. 2016;4:38.
37. Sekiya I, et al. Adipogenic differentiation of human adult stem cells from bone marrow stroma (MSCs). J Bone Miner Res. 2004;19(2):256–64.
38. Kilroy G, et al. Isolation of murine adipose-derived stromal/stem cells for adipogenic differentiation or flow cytometry-based analysis. Methods Mol Biol. 2018;1773:137–46.
39. Jaiswal N, et al. Osteogenic differentiation of purified, culture-expanded human mesenchymal stem cells in vitro. J Cell Biochem. 1997;64(2):295–312.
40. Song IH, Caplan AI, Dennis JE. In vitro dexamethasone pretreatment enhances bone formation of human mesenchymal stem cells in vivo. J Orthop Res. 2009;27(7):916–21.

41. Barry F, et al. Chondrogenic differentiation of mesenchymal stem cells from bone marrow: differentiation-dependent gene expression of matrix components. Exp Cell Res. 2001;268(2):189–200.
42. Mauck R, Yuan X, Tuan R. Chondrogenic differentiation and functional maturation of bovine mesenchymal stem cells in long-term agarose culture. Osteoarthr Cartil. 2006;14(2):179–89.
43. Fan L, et al. Enhancement of the chondrogenic differentiation of mesenchymal stem cells and cartilage repair by ghrelin. J Orthop Res. 2019;37(6):1387–97.
44. Ghaben AL, Scherer PE. Adipogenesis and metabolic health. Nat Rev Mol Cell Biol. 2019;20(4):242–58.
45. Villarroya F, et al. Brown adipose tissue as a secretory organ. Nat Rev Endocrinol. 2017;13(1):26.
46. Smitka K, Marešová D. Adipose tissue as an endocrine organ: an update on pro-inflammatory and anti-inflammatory microenvironment. Prague Med Rep. 2015;116(2):87–111.
47. Churm R, et al. Ghrelin function in human obesity and type 2 diabetes: a concise review. Obes Rev. 2017;18(2):140–8.
48. Gustafson B, et al. Insulin resistance and impaired adipogenesis. Trends Endocrinol Metab. 2015;26(4):193–200.
49. Reisin E, Owen J. Treatment: special conditions: metabolic syndrome: obesity and the hypertension connection. J Am Soc Hypertens. 2015;9(2):156–9.
50. Chappell VA, et al. Tetrabromobisphenol-A promotes early adipogenesis and lipogenesis in 3T3-L1 cells. Toxicol Sci. 2018;166(2):332–44.
51. Tung Y-C, et al. Cellular models for the evaluation of the antiobesity effect of selected phytochemicals from food and herbs. J Food Drug Anal. 2017;25(1):100–10.
52. De Sa PM, et al. Transcriptional regulation of adipogenesis. Compr Physiol. 2017;7:635–74.
53. Tencerova M, Kassem M. The bone marrow-derived stromal cells: commitment and regulation of adipogenesis. Front Endocrinol. 2016;7:127.
54. Rony RIK, et al. Differential expression of PPARγ and CHOP-10 during Adipogenic differentiation of human bone marrow derived mesenchymal stem cells. FASEB J. 2018;32(1_suppl):lb17.
55. Fu M, et al. A nuclear receptor atlas: 3T3-L1 adipogenesis. Mol Endocrinol. 2005;19(10):2437–50.
56. Ruiz-Ojeda F, et al. Cell models and their application for studying adipogenic differentiation in relation to obesity: a review. Int J Mol Sci. 2016;17(7):1040.
57. Forni MF, et al. Murine mesenchymal stem cell commitment to differentiation is regulated by mitochondrial dynamics. Stem Cells. 2016;34(3):743–55.
58. Moreno-Navarrete JM, Fernández-Real JM. Adipocyte differentiation. In: Adipose tissue biology. New York: Springer; 2017. p. 69–90.
59. Yuan Z, et al. PPARγ and Wnt signaling in adipogenic and osteogenic differentiation of mesenchymal stem cells. Curr Stem Cell Res Ther. 2016;11(3):216–25.
60. Bennett CN, et al. Regulation of Wnt signaling during adipogenesis. J Biol Chem. 2002;277(34):30998–1004.
61. Gross B, et al. PPARs in obesity-induced T2DM, dyslipidaemia and NAFLD. Nat Rev Endocrinol. 2017;13(1):36.
62. Salazar-Roa M, Malumbres M. Fueling the cell division cycle. Trends Cell Biol. 2017;27(1):69–81.
63. Chiurchiù V, Leuti A, Maccarrone M. Bioactive lipids and chronic inflammation: managing the fire within. Front Immunol. 2018;9:38.
64. Nagao K, Yanagita T. Bioactive lipids in metabolic syndrome. Prog Lipid Res. 2008;47(2):127–46.
65. Fahy E, et al. Update of the LIPID MAPS comprehensive classification system for lipids. J Lipid Res. 2009;50(Suppl):S9–S14.
66. Fahy E, et al. A comprehensive classification system for lipids. Eur J Lipid Sci Technol. 2005;107(5):337–64.

67. Holm R. Bridging the gaps between academic research and industrial product developments of lipid-based formulations. Adv Drug Deliv Rev, 2019. https://doi.org/10.1016/j.addr.2019.01.009.
68. Jones SF, Infante JR. Molecular pathways: fatty acid synthase. Clin Cancer Res. 2015;21(24):5434–8.
69. Hashimoto M, Hossain S. Fatty acids: from membrane ingredients to signaling molecules, in biochemistry and health benefits of fatty acids. London: IntechOpen; 2018.
70. Han X, Zhou Y. Application of lipidomics in nutrition research. In: Metabolomics as a tool in nutrition research. Amsterdam: Elsevier; 2015. p. 63–84.
71. Matsumaru T, et al. Synthesis of glycerolipids containing simple linear acyl chains or aromatic rings and evaluation of their Mincle signaling activity. Chem Commun. 2019;55(5):711–4.
72. Khoury S, et al. Quantification of lipids: model, reality, and compromise. Biomol Ther. 2018;8(4):174.
73. Lingwood D, Simons K. Lipid rafts as a membrane-organizing principle. Science. 2010;327(5961):46–50.
74. Merrill AH Jr. Sphingolipids. In: Biochemistry of lipids, lipoproteins and membranes. Amsterdam: Elsevier; 2008. p. 363–97.
75. Chun J, Hartung H-P. Mechanism of action of oral fingolimod (FTY720) in multiple sclerosis. Clin Neuropharmacol. 2010;33(2):91–101.
76. Dickson RC, Lester RL. Sphingolipid functions in Saccharomyces cerevisiae. Biochim Biophys Acta. 2002;1583(1):13–25.
77. Fahy E, et al. A comprehensive classification system for lipids. J Lipid Res. 2005;46:839–61.
78. Raetz CR, et al. Discovery of new biosynthetic pathways: the lipid A story. J Lipid Res. 2009;50(Suppl):S103–8.
79. Pfeifer BA, Khosla C. Biosynthesis of polyketides in heterologous hosts. Microbiol Mol Biol Rev. 2001;65(1):106–18.
80. Lim Y, Go M, Yew W. Exploiting the biosynthetic potential of type III polyketide synthases. Molecules. 2016;21(6):806.
81. Demel RA, De Kruyff B. The function of sterols in membranes. Biochim Biophys Acta. 1976;457(2):109–32.
82. Wolstenholme GEW, O'Connor CM. Quinones in electron transport, vol. 947. Hoboken: Wiley; 2009.
83. Lydic TA, Goo Y-H. Lipidomics unveils the complexity of the lipidome in metabolic diseases. Clin Transl Med. 2018;7(1):4–4.
84. Zhao YY, et al. Lipidomics applications for disease biomarker discovery in mammal models. Biomark Med. 2015;9(2):153–68.
85. Zhao YY, Cheng XL, Lin RC. Lipidomics applications for discovering biomarkers of diseases in clinical chemistry. Int Rev Cell Mol Biol. 2014;313:1–26.
86. Liaw L, et al. Lipid profiling of in vitro cell models of adipogenic differentiation: relationships with mouse adipose tissues. J Cell Biochem. 2016;117(9):2182–93.
87. Yang J-Y, et al. Regulation of adipogenesis by medium-chain fatty acids in the absence of hormonal cocktail. J Nutr Biochem. 2009;20(7):537–43.
88. Kim H-K, et al. Docosahexaenoic acid inhibits adipocyte differentiation and induces apoptosis in 3T3-L1 preadipocytes. J Nutr. 2006;136(12):2965–9.
89. Dwyer JR, et al. Oxysterols are novel activators of the hedgehog signaling pathway in pluripotent mesenchymal cells. J Biol Chem. 2007;282(12):8959–68.
90. Eaton S. Multiple roles for lipids in the hedgehog signalling pathway. Nat Rev Mol Cell Biol. 2008;9(6):437.
91. Gregg EW, Shaw JE. Global health effects of overweight and obesity. N Engl J Med. 2017;377(1):80–1.
92. Dixon J. The global burden of obesity and diabetes. In: Minimally invasive bariatric surgery. New York: Springer; 2015. p. 1–6.

93. Hossain P, Kawar B, El Nahas M. Obesity and diabetes in the developing world—a growing challenge. N Engl J Med. 2007;356(3):213–5.
94. Low S, Chin MC, Deurenberg-Yap M. Review on epidemic of obesity. Ann Acad Med Singap. 2009;38(1):57.
95. Lavie CJ, et al. Management of cardiovascular diseases in patients with obesity. Nat Rev Cardiol. 2018;15(1):45.
96. Hurt RT, et al. Obesity epidemic: overview, pathophysiology, and the intensive care unit conundrum. J Parenter Enter Nutr. 2011;35(5_suppl):4S–13S.
97. Arnold M, et al. Obesity and cancer: an update of the global impact. Cancer Epidemiol. 2016;41:8–15.
98. Collaborators GO. Health effects of overweight and obesity in 195 countries over 25 years. N Engl J Med. 2017;377(1):13–27.
99. Rennie K, Jebb S. Prevalence of obesity in Great Britain. Obes Rev. 2005;6(1):11–2.
100. Husky MM, et al. Differential associations between excess body weight and psychiatric disorders in men and women. J Women's Health. 2018;27(2):183–90.
101. Webb P, et al. Hunger and malnutrition in the 21st century. BMJ. 2018;361:k2238.
102. Jacob CS, de Alba Carolina T. An evidence-based review of dietary supplements on inflammatory biomarkers in obesity. Curr Res Nutr Food Sci J. 2018;6(2):284–93.
103. Khan M. Complications of cryolipolysis: paradoxical adipose hyperplasia (PAH) and beyond. Aesthet Surg J. 2019;39(8):NP334–42.
104. Considine RV, et al. Paracrine stimulation of preadipocyte-enriched cell cultures by mature adipocytes. Am J Physiol Endocrinol Metab. 1996;270(5):E895–9.
105. Tang QQ, Lane MD. Adipogenesis: from stem cell to adipocyte. Annu Rev Biochem. 2012;81:715–36.
106. Cook D, Genever P. Regulation of mesenchymal stem cell differentiation, in transcriptional and translational regulation of stem cells. Adv Exp Med Biol. 2013;786:213–29.
107. Lindroos B, Suuronen R, Miettinen S. The potential of adipose stem cells in regenerative medicine. Stem Cell Rev Rep. 2011;7(2):269–91.
108. Ong WK, Sugii S. Adipose-derived stem cells: fatty potentials for therapy. Int J Biochem Cell Biol. 2013;45(6):1083–6.
109. Coelho M, Oliveira T, Fernandes R. Biochemistry of adipose tissue: an endocrine organ. Arch Med Sci. 2013;9(2):191.
110. Galic S, Oakhill JS, Steinberg GR. Adipose tissue as an endocrine organ. Mol Cell Endocrinol. 2010;316(2):129–39.
111. Jo J, et al. Hypertrophy and/or hyperplasia: dynamics of adipose tissue growth. PLoS Comput Biol. 2009;5(3):e1000324.
112. Payab M, et al. Stem cell and obesity: current state and future perspective. Adv Exp Med Biol. 2018;1089:1–22.
113. Joe AW, et al. Depot-specific differences in adipogenic progenitor abundance and proliferative response to high-fat diet. Stem Cells. 2009;27(10):2563–70.
114. Matsushita K, Dzau VJ. Mesenchymal stem cells in obesity: insights for translational applications. Lab Investig. 2017;97(10):1158.
115. Niemelä S, et al. Adipose tissue and adipocyte differentiation: molecular and cellular aspects and tissue engineering applications. Top Tissue Eng. 2008;4(1):26.
116. Cleal L, Aldea T, Chau Y-Y. Fifty shades of white: understanding heterogeneity in white adipose stem cells. Adipocytes. 2017;6(3):205–16.
117. Cawthorn WP, Scheller EL, MacDougald OA. Adipose tissue stem cells meet preadipocyte commitment: going back to the future. J Lipid Res. 2012;53(2):227–46.
118. Tang W, et al. White fat progenitor cells reside in the adipose vasculature. Science. 2008;322(5901):583–6.
119. Matsushita K. Mesenchymal stem cells and metabolic syndrome: current understanding and potential clinical implications. Stem Cells Int. 2016;2016:2892840.

120. Poulos SP, et al. The increasingly complex regulation of adipocyte differentiation. Exp Biol Med. 2016;241(5):449–56.
121. Moseti D, Regassa A, Kim W-K. Molecular regulation of adipogenesis and potential anti-adipogenic bioactive molecules. Int J Mol Sci. 2016;17(1):124.
122. Postle AD. Lipidomics. Curr Opin Clin Nutr Metab Care. 2012;15(2):127–33.
123. Murphy SA, Nicolaou A. Lipidomics applications in health, disease and nutrition research. Mol Nutr Food Res. 2013;57(8):1336–46.
124. Yang K, Han X. Lipidomics: techniques, applications, and outcomes related to biomedical sciences. Trends Biochem Sci. 2016;41(11):954–69.
125. Han X. Lipidomics for studying metabolism. Nat Rev Endocrinol. 2016;12(11):668.
126. Cristancho AG, Lazar MA. Forming functional fat: a growing understanding of adipocyte differentiation. Nat Rev Mol Cell Biol. 2011;12(11):722.
127. Lapid K, Graff JM. Form(ul)ation of adipocytes by lipids. Adipocytes. 2017;6(3):176–86.
128. de Ferranti S, Mozaffarian D. The perfect storm: obesity, adipocyte dysfunction, and metabolic consequences. Clin Chem. 2008;54(6):945–55.
129. Hammarstedt A, et al. Impaired adipogenesis and dysfunctional adipose tissue in human hypertrophic obesity. Physiol Rev. 2018;98(4):1911–41.
130. Han X. An update on lipidomics: progress and application in biomarker and drug development. Curr Opin Mol Ther. 2007;9(6):586–91.
131. Haraszti RA, et al. High-resolution proteomic and lipidomic analysis of exosomes and microvesicles from different cell sources. J Extracell Vesicles. 2016;5(1):32570.
132. Nguyen A, et al. Using lipidomics analysis to determine signalling and metabolic changes in cells. Curr Opin Biotechnol. 2017;43:96–103.
133. Qian S-W, et al. Characterization of adipocyte differentiation from human mesenchymal stem cells in bone marrow. BMC Dev Biol. 2010;10:47.
134. Mohammadi Z, et al. Differentiation of adipocytes and osteocytes from human adipose and placental mesenchymal stem cells. Iran J Basic Med Sci. 2015;18(3):259–66.
135. Marquez MP, et al. The role of cellular proliferation in adipogenic differentiation of human adipose tissue-derived mesenchymal stem cells. Stem Cells Dev. 2017;26(21):1578–95.
136. Montacir H, et al. N-glycosylation profile of undifferentiated and adipogenically differentiated human bone marrow Mesenchymal stem cells: towards a next generation of stem cell markers. Stem Cells Dev. 2013;22(23):3100–13.
137. Sarantopoulos CN, et al. Elucidating the preadipocyte and its role in adipocyte formation: a comprehensive review. Stem Cell Rev. 2018;14(1):27–42.
138. Masoodi M, et al. Lipid signaling in adipose tissue: connecting inflammation & metabolism. Biochim Biophys Acta. 2015;1851(4):503–18.
139. Lee Y-H, Mottillo EP, Granneman JG. Adipose tissue plasticity from WAT to BAT and in between. Biochim Biophys Acta. 2014;1842(3):358–69.
140. Titz B, et al. Proteomics and lipidomics in inflammatory bowel disease research: from mechanistic insights to biomarker identification. Int J Mol Sci. 2018;19(9):2775.

OMICs Profiling of Cancer Cells

Bagher Larijani, Parisa Goodarzi, Motahareh Sheikh Hosseini, Solmaz M. Nejad, Sepideh Alavi-Moghadam, Masoumeh Sarvari, Mina Abedi, Maryam Arabi, Fakher Rahim, Najmeh Foroughi Heravani, Mahdieh Hadavandkhani, and Moloud Payab

Abstract The number of people survive from cancer is increasing in the USA due to the advances in the early detection and treatment. Cancer is a complex disease caused by several factors such as genetics, epigenetics, proteomics, and transcriptional alterations or cellular damage that is resulted from several factors through genetic mutations and environmental effects. Early diagnosis has a pivotal role in the treatment or improving outcomes of cancer. Therefore, detecting cancer at early stages is a key challenge in cancer medicine and increases the survival rate. For early diagnosis, some genetics, proteomics, and metabolomics profiling should be considered using OMICs technologies. Traditional technologies using simplistic approach such as chemotherapy and surgery are relatively insufficient to facing challenges in the treatment of cancer. As a result, OMICs technologies mainly focus

B. Larijani
Endocrinology and Metabolism Research Center,
Endocrinology and Metabolism Clinical Sciences Institute,
Tehran University of Medical Sciences, Tehran, Iran

P. Goodarzi
Brain and Spinal Cord Injury Research Center, Neuroscience Institute,
Tehran University of Medical Sciences, Tehran, Iran

M. Sheikh Hosseini · S. M. Nejad · M. Sarvari
Metabolomics and Genomics Research Center, Endocrinology and Metabolism
Molecular-Cellular Sciences Institute, Tehran University of Medical Sciences, Tehran, Iran

S. Alavi-Moghadam · M. Abedi · M. Arabi · N. Foroughi Heravani
M. Hadavandkhani
Cell Therapy and Regenerative Medicine Research Center, Endocrinology and Metabolism
Molecular-Cellular Sciences Institute, Tehran University of Medical Sciences, Tehran, Iran

F. Rahim
Health Research Institute, Thalassemia and Hemoglobinopathies Research Center,
Ahvaz Jundishapur University of Medical Sciences, Ahvaz, Iran

M. Payab (✉)
Obesity and Eating Habits Research Center, Endocrinology and Metabolism
Molecular-Cellular Sciences Institute, Tehran University of Medical Sciences, Tehran, Iran

B. Arjmand (ed.), *Genomics, Proteomics, and Metabolomics*, Stem Cell Biology and Regenerative Medicine, https://doi.org/10.1007/978-3-030-27727-7_8

on the detection of entire genes which is applied into genomics, mRNA (transcriptomics), proteins (proteomics), metabolites (metabolomics), and lipids (lipidomics) in cells. The term genomics refers to the study of structure and function of DNA. Gene expression studies which is referred to transcriptomics are one of the oldest OMICs technologies, as it is the analysis of the entire RNA sequences in a cell. Proteomics technologies can identify the protein changes caused by disease process. Metabolomics is the study of small molecules, which are metabolites and are found in cells, tissues, and bio-fluids of an organism. Patterns of plasma lipid opulence are referred to as the lipidome. OMICs technologies, which system biology bring, are valuable tools for comprehensive analysis. The availability of DNA sequencing automatically enabled the sequencing of genomes; immunohistochemistry, which is one of the protein-based histopathological assays, has been the traditional basis of laboratory-based tumor characterization. Microarray and mass spectrometry analysis enabled comprehensive transcriptional profiling and lead to large-scale proteomics and metabolomics analysis. Scientists hope that with future analyzing of OMICs data, we can increase our therapeutic productivity for molecular targets of cancer therapies. The data of cancer OMICs are rapidly collected and provided an invaluable resource for identifying novel targets in the treatment of cancer, and will accelerate with developed diagnostic technologies and advanced novel methods in near future.

Keywords Biomarkers · Cancer · Early detection · OMICs technology · Treatment associated

1 Introduction

In 2019, the estimated numbers of deaths due to cancers in the USA are almost 1700 deaths per day. Of course, in other parts of the world the mortality rates of cancer may become different depending on various internal factors such as inherited mutations and lifestyles [1]. The most frequent lethal cancers among Americans are lung, prostate, and colorectal cancers in men and the lung, breast, and the colorectal cancers in women. Meanwhile, one-third of all cancer deaths are from lung cancer [2]. On the other hand, the number of people survive from cancer is increasing in the USA due to the advances in early detection and treatment [3]. Cancer is a complex disease caused by variety of factors such as genetics, epigenetics, proteomics, and transcriptional variations or cellular damage that is resulted from several items through genetic mutations and environmental effects [3–5]. In addition to the technical challenges of somatic variants, most genetic changes are benign and do not cause growth of cancer cells. Therefore, determining the major driver mutations and their signaling pathways that lead to cancer is a major challenge. On the other hand, although some cancers share common genetic signatures throughout individuals, there is still a significant difference in the levels of driver mutations, which can lead to individualized approach to prognosis and treatments. Indeed, many of the strategies such as network methods will be effective in identifying genetic mechanisms of cancers [4, 6]. Cancer diagnosis and effective treatment especially in symptomatic patients are timely and crucial. Additionally, early diagnosis has an important role in treatment or improving outcomes of cancer [1]. For early diagnosis, some genetics, proteomics, or metabolomics profiling should be considered using OMICs technologies. Some of the measurable indicators which play an important role in cancer diagnosis and also are specific to any disease or cancer are introduced as biomarkers. Detecting of specific biomarkers is an alternative approach to early diagnosis, staging, and management of cancer. Recently, the detection of circulating tumor cells (CTCs) as a new method is considered seriously [7]. Cancer cells have different metabolisms in comparison with normal cells; therefore, their metabolomics profiles are more specific according to staging and management of cancer. Proteomics plays a pivotal role in the discovery of cancer biomarkers [6]. As a result, early diagnosis, management of cancers, gene regulation, and gene editing are greatly facilitated by OMICs (genomics, transcriptomics, proteomics, and metabolomics) technologies. Genomics analysis offers a clue to gene regulation and cancer management. OMICs technologies in cancer management have some reasonable advantages including potential identification of the large-scale changes for achieving more comprehensive huge data analysis compare to thousands of individual tests that provides more feasible and cheaper methods. The emerging technologies of OMICs are used for cancer and drug investigations, in combination with bioinformatics and computer-based modeling techniques. Recently, by advances in OMICs, new opportunities in molecular assays have been administered that may result in cancer diagnosis and treatment [4, 6]. With the advancement of OMICs technology and data analysis, detailed molecular information can be gathered from diseased cells. Following the mentioned advancement, a deeper biological insight into the incurable diseases such as cancers can be provided [8, 9]. In summary, OMICs will increase our knowledge of molecular characteristics of cancers in both qualitative and quantitative forms. In addition, OMICs can weaponize scientists to identify, characteristic, classify cancers in the levels of gene, protein, and metabolite and their signaling pathways to achieve more effective diagnosis and management methods. In this chapter, we will describe new research opportunities using OMICs technologies information in new therapies [4, 6].

2 Cancer and Cancer Cells

As cancer is a heterogeneous disease, so analyzing cell to cell variation is necessary to understand tumor development, metastasis, and therapeutic responses. Therefore, at present there are extensive research on hallmarks in cancer that leads to invasion and metastasis [10]. Detecting cancer at early stages is a key challenge in cancer medicine and increases the survival rate. For example, in the metastasized cases of breast cancer, the survival rate is 23%, while the survival rate increases to 97% when it is detected at early stages where malignancy is maintained in the origin organ. Thus, for diagnosis of cancer at early stages, researchers need specific biomarkers [11]. There are three main groups of biomarkers for cancer diseases including diagnostic, prognostic, and predictive biomarkers. The diagnostic biomarkers detect cancer at early stages and in noninvasive way; predictive biomarkers indicate the patient's response to a targeted therapy, and prognostic biomarkers provide information of the natural course of a disease. Analysis of biomarkers in recent decades has found a new method for pathological analysis and developed a new insight into individualized medicine [12, 13]. The critical therapeutic target in tumors is cancer stem cells because of their responsibility in tumor repopulation after treatment.

Normal and cancer cells can be distinguished by single cell analysis (SCA) at different stages. The cancer stem cells are potent to detect a small proportion of the whole tumor but they are responsible for developing tumors. The SCA has the potential to detect cancer stem cells and their unique ability. The advancement of SCA can accelerate researching related to different biological aspects, diagnostics, and also genetic diagnosis [14].

3 Monitoring of Cancer Prognosis, Diagnosis, and Cancer Treatment Options

The history of cancer treatment reaches back to the eighteenth century, when Scot John Hunter, who is one of the initiators of modern surgery, disclosed that if a tumor had not invaded to neighboring tissues, it should be removed. At 1900, small doses of radiation were used in the treatment of skin cancer [15].

Then after, the first chemotherapy drug was accepted in 1949 by US Food and Drug Administration (FDA). This drug had been designed based on a poison gas used in the First World War. During 1950s, some drugs such as methotrexate were used to cure metastatic cancers. Then by combination therapy, several drugs were used at once for treatment of leukemia and lymphoma. At 1970s, by progression of imaging techniques, surgery and radiation methods were improved. In the late 1990s, drugs based on monoclonal antibodies had been licensed [15].

In the twenty-first century, due to the impressive prevalence of various types of cancer, our knowledge of cancer biology has greatly improved that consequently resulted in advancement of cancer treatment methods. Cancer can be treated by various methods such as surgery, chemotherapy, hormone therapy, radiotherapy, targeted therapy, and immunotherapy [16]. In the early 2000s, some drugs by the characterization of interrupting the tumor signaling pathways opened new windows in cancer treatment for researchers.

The perfect treatment in cancer is removing the cancerous tissue/organ with no loss or adverse effect on the other parts of the body. During the last decade, the cancer treatment approaches have greatly changed by improvements in molecular techniques and tumor biology science. Previously, cancer was categorized and treated according to the organs of origin or simplistic histomorphologic features.

Recent studies cleared that the development of molecularly targeted therapeutic methods is required for cancer treatment [17].

Currently, cancer patients are treated by various methods such as surgery, chemotherapy, radiotherapy, hormonal therapy, and targeted therapy. However, advanced methods suggest addition of new technology for molecular profiling of tumors and finding predictive molecular objects. Choosing therapy method depends on the size and location of tumor, the stage of cancer, age and status of per patient [18].

3.1 Surgery

In cancer treatment, different types of surgery are used. These types include curative, preventive, diagnostic, palliative, and debulking surgery [19].

The curative surgery is a primary treatment that removes the localized tumor from body. *Preventive surgery* removes not cancerous tissues and prevents the development of malignancy. By *diagnostic surgery*, the tissue samples for pathological studies are obtained. The *palliative surgery* is a method in the treatment of advanced stages of disease, and *debulking surgery* removes a part of tumor if removing the entire tumor may damage the organs [20].

In cancer treatment, various techniques such as cryosurgery, lasers, hyperthermia, and photodynamic therapy have also been used to destroy the malignant tissues [21].

3.2 Radiation Therapy (Radiotherapy)

In this method, high doses of radiation have been prescribed to kill cancer cells and tumors or slowing down the tumor cells promotion. Radiation causes damage in DNA structure which results in inhibition of cell division [22].

3.3 Chemotherapy

The aims of chemotherapy are both cure the disease and reduction of symptoms. All the alkylating agents, antimetabolites, plant derived compounds, topoisomerase inhibitors, and cytotoxic antibiotics are various types of chemotherapeutics. These drugs depending on their type have side effects. Mostly, they have cytotoxic effects on growth of cell division especially in blood cells or intestine cells. The chemotherapeutic induced toxicities may reveal shortly after administration or chronically [23].

3.4 Immunotherapy

During cancer, the natural defense system of the body has been suppressed by various causes. Immune therapy is a treatment method that improves immune system and helps body to fight with cancer by slowing the growth of tumor or stopping invasion of malignancy to nearby tissues, as well. Immunotherapy has several types, such as checkpoint inhibitors, adoptive cell transfers, monoclonal antibodies, vaccines, and cytokines [24].

The main purpose of diagnosis and treatment procedure in patients suffer from cancer is cure and prolonging their life. Providing standard and equitable ways for accurate diagnosis and correct detection based on evidences in patient will help physicians to design effective treatment programs based on staging condition of patient. Per year, patients suffering from cancer are subjected to various types of costly tests for correct analysis of their disease. Several types of diagnostic tests are listed as below:

- Pathologic tests that include microscopic assessment of abnormal cells.
- Imaging tests that include visualization of abnormalities and malignancies. X-rays, computed tomography scan (CT scan) and magnetic resonance imaging (MRI).
- Blood tests, which by measurement and analyzing the blood substances, indicate progression or remission of disease.
- Measurement of tumor markers in blood, urine, or other bio-fluids and tissues.
- OMICs technologies including special laboratory evaluation of DNA, RNA, protein, and other metabolites of the body which identify cellular genomic and abnormalities [25].

Although these tests are named "diagnostic," they are used for monitoring the progression of cancer. That is to say, these tests show the activation, progression, and response to treating drugs. Another important role of diagnosis test is in screening individuals who are at risk. Identification of genetic and cellular abnormalities in asymptomatic individuals helps physicians to be aware of the increasing risk of disease in healthy person [25].

4 Genetics of Cancer Cells

In 2001, human genome sequencing which completed by the human genome project was conducted by the extensive advancement of high DNA sequencing technologies. After this advancement, scientists expected that they can unveil the molecular reasons which lead to different diseases including diabetes, hypertension, infectious diseases, and more importantly cancer. On the other hand, for developing cancer, several changes accumulated in the genome and resulted in migration, invasion, and accelerated growth. In addition, deregulation of cell growth is caused by the epigenetic mechanisms such as DNA methylation, chromatin modifications, histone acetylation, and gene silencing. Therefore, pathophysiology of diseases particularly cancer or finding vigilant location of mutated genes especially when the mutation is silent cannot be understood by simply gene sequencing. Each group of cancer cells defined by different combination of genetic changes (mutations, deletion, etc.) and the process of tumor generation is different dependent on the type of genetic lesions [10]. The emerging of transcriptome analysis has facilitated the realizing of alterations of gene expression in cancerous cells. Furthermore, in the field of cancer, understanding of the relationship between the mRNA levels and the content of protein is so important because all the mRNA molecules do not translate to the proteins. In other words, finding the regular correlation between mRNA levels and protein content is so important for transcriptional factors and metabolic enzymes. This information can be realized by proteomics analysis, for example, evaluating the variation in the protein content and protein activity. Interpreting data that extracted from transcriptomics, genomics, and proteomics illustrates important information. For example, the data show the relationship between specific mutation and loss or gain of function in the mRNA or effective protein content [10].

As a result, the OMIICs technologies play a critical role in the management of cancer development [10]. OMICs technologies including genomics, epigenetics, transcriptomics, metabolomics, proteomics, lipidomics, and interactomics are the studies of genes, proteins, lipids, and other metabolites in large scales. With the advancement of OMICs studies, we can identify and quantify the whole components of cells and also the network and intracellular pathway. The detecting processes of OMICs technology in genomics, epigenomics, proteomics, and metabolomics are microarrays, DNA sequencing, protein arrays, mass spectrometry (MS), and nuclear magnetic resonance spectrometry (NMR) [14].

5 Biomarkers in Cancer

Over the past few years, by the advancement of precise medical treatment and early diagnosis of cancer, the percentage of cancer mortality has been decreased; one of the most important factors in this field is the use of biomarkers in cancer staging and personalized medicine which improve the survival rate in cancer. Exploration of early signs of the disease with new technologies has a key role in cancer management. Detecting tumor biomarkers cover a broad range from nucleic acids, proteins, small metabolites, the cytogenetic and cytokinetic parameters to the entire tumor cells found in the body fluid. Today, with the development of cancer diagnostic methods, the presence of cancer in the patient can be detected by cancer biomarkers. Hence, biomarkers are used to detect known tumors. The biomarkers are produced by the body in response to cancer or by the tumor itself. Different type of biomarkers such as proteomic, genetic, epigenetic, and imaging biomarkers have a pivotal role in cancer diagnosis and prognosis. Table 1 shows the examples of various molecular cancer biomarkers [7, 33, 34].

Table 1 Various molecular cancer biomarkers

Cancer types	Biomarker	Biomarker types	Reference
Breast	ER∗/PR∗ HER-2∗	Prognostic Treatment	[26]
Colorectal	EGFR∗/KRAS∗ UGT1A1∗	Predictive Diagnostic	[27]
Gastric	HER-2∗	Treatment	[28]
Leukemia–lymphoma	CD20∗ CD30∗	Diagnostic, treatment Predictive	[29] [30]
Lung	EGFR∗ KRAS∗	Predictive	[31]
Melanoma	BRAF∗	Prognostic and predictive	[32]

ER∗\PR∗ Estrogen Receptor/Progesterone Receptor, *HER2∗* Human Epidermal Growth Factor Receptor 2, *EGFR∗* Epidermal Growth Factor Receptor, *KRAS∗* Oncogene in Kirsten RAt Sarcoma Virus, *UGT1A1∗* Bilirubin Uridine Diphosphate GlucuronosylTransferase (bilirubin-UGT), *CD20∗–CD30∗* Cluster of Differentiation, *BRAF∗* Murine Sarcoma Viral Oncogene Homolog B

6 OMICs Technology (Fig. 1)

6.1 Genomics

Genomics focuses on the structure, function, activity, design, and alteration of genome, which indicates the study of individual genes and the role of these genes in inheritance. The main goal of genomics is the definition and quantified assessment of genes, which address the construction of proteins with the enzymes cooperation and messenger molecules. This technique is also associated with the genome sequencing and genome analysis using high-throughput DNA sequencing and bioinformatics sciences. Improvements in genomics have generated a great revolution in biologic research and facilitated discovery of some biological systems such as brain or tumors, which are complex systems. The acquisition of mutations influences the result in tumor phenotype as therefore, sequencing of cancer genomes increase our insight into the understanding of driver mutations [36].

Genomics also is beneficial in diagnosis, prognosis, and treatment of cancer. Single-cell genomics in the tumor research facilitated the detection of tumors at early stages. Moreover, with the heterogeneity of cancer, the requirement of precise medicine is a critical issue in oncology. Therefore, cancer genome study can analyze genomic DNA in millions of cells and accumulating and representing data across population and computational approaches and interpret phylogenetic trees for each tumor. As a result, this information in genomics can increase our knowledge about how tumors evolve during progression and under the pressure of therapy [35, 36].

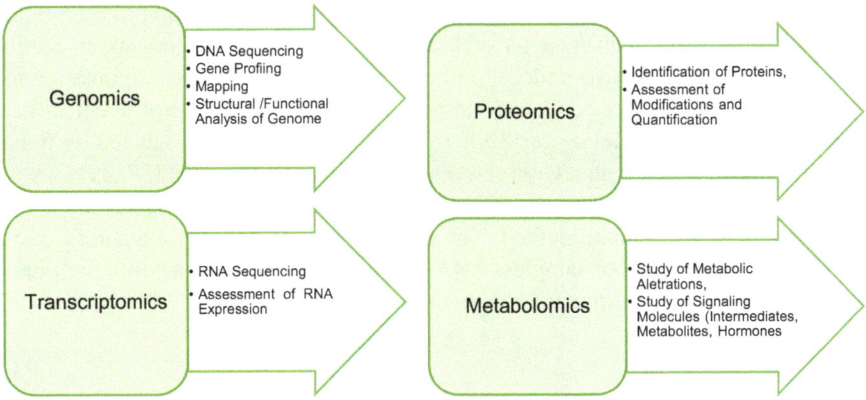

Fig. 1 A simple view of OMICS technologies. A simple view of OMICS technologies. OMICs technologies including genomics (DNA sequencing, gene profiling, mapping, structural /functional analysis of genome), proteomics (identification of proteins, assessment of modifications and quantification), metabolomics (study of metabolic alterations, study of signaling molecules (intermediates, metabolites, hormones)), and transcriptomics (RNA sequencing, assessment of RNA expression) have changed the view of medical researches

6.2 Transcriptomics

Transcripts can influence cell structure and effect on genes regulating. Analysis of the whole RNA sequences in a cell is called transcriptomics, and with this technique researchers are potent to determine turn off and turn on the gene in cells or tissues of an organism. Also, in a certain cell or tissue, counting the number of transcripts is possible to determine expression of genes.

Although in all organisms, cells have genetic similarities, but different cells have various models of gene expression. However, these diversities are responsible for various functional and behavioral specialties of cells and tissues, which is valuable in health and disease. Analyzing transcriptomes in various cell types gives a deeper understanding of cell function, gene activity, and the effect of these factors in a disease process.

Due to multiple factors, carcinogenesis still is unrevealed because various determinants such as genetic and epigenetic modifications play role in this process. In other words, cancer is an intricate process that begins with genetic alterations, including mutations, and ends with proteomic changes that leads to the uncontrollable cellular proliferation. In the next step, the differentiated cancerous cells invade to other tissues and metastasis is the final point.

In cancer studies, the transcriptomic data show that gene expression is higher in cancerous cells compared with healthy cells. The transcriptomic data also presents a good perspective to design a cancer research because it provides the ability to study cells from genetic to proteomic level and find biomarkers and or new targets for treatment.

Microarrays and next-generation sequencing technologies (NGS) are good examples of transcriptomic techniques. Additionally, transcription can be studied at the level of individual cells called single-cell transcriptomics.

Microarrays: Microarray is an experimental tool for the detection of transcriptomes and has some requirements such as genomic DNA of the organism. Some molecules, which are also called "probes," are short nucleotide oligomers and typically are arrayed in a conducted part on a glass slide. These probes detect the genetic expression of messenger RNA transcripts and have been labeled by fluorescent. The achieved data are observed as high-resolution images [37–39].

RNA-Seq: RNA-Seq is the consolidation of a high-throughput sequencing technique and a computational method. The aim of this technique is to quantify transcripts in RNA extracts of samples. RNA-Seq identifies genes at a particular point of genome in a specific time and is able to produce plenty of short DNA [40].

6.3 Proteomics

Proteins are vital molecules in living organisms. Proteomics technologies are the study of proteins or the proteome of the cell and can identify the protein changes caused by disease process [41]. Proteomics analyze the entire set of proteins. The foundation of traditional proteomics data analysis in the laboratory was protein-based histopathological assays based on tumor specification using immunohistochemistry but the data obtained from this information were insufficient.

In recent years, cancer is become a great health challenge despite the development of cancer diagnosis and treatment. The emergence of novel technologies increases the pace of cancer research and has a huge impact on the care of cancer patients. These developments have already demonstrated the power of molecular analysis in cancer management. With these advancements in molecular medicine, identifying genetic events involved in tumor progression has been facilitated. As a result, new technologies in molecular analysis revolutionized the analysis of healthy and diseased cells. The development of genomics and proteomics technologies hoped to provide the advance knowledge in treatment likewise in discovery and determent. Therefore, developing OMICs technology based on proteomics facilitated analyzing the full set of proteins expressed by tumor cells [42, 43].

6.4 Metabolomics

As cancer is a metabolic disease, metabolic studies are important in discovery and treatment of it. Metabolomics is the study of small molecules which are metabolites and are found in cells, tissues, and bio-fluids of an organism [44, 45].

In recent years, scientists increase their attention on early detection of cancer with defining metabolic alteration during carcinogenesis and find some useful treatment methods in cancer diagnosis at early stages. This revolution in therapeutic approaches is due to the development in the accessibility of metabolomics data and increases the detection of beneficial cancer metabolite biomarkers and the discovery of oncometabolites using metabolomics. The history of this new findings reaches back to 80 years ago when the relation between cancer and metabolic alteration has been discovered [46, 47]. With the development of methods and statistical analysis of cancer diagnostic including analyzing the metabolic changes in cancer, scientists have found deeper understanding of cancer metabolites via metabolomics. Therefore, we can increase our recognition of how cells use glycolysis extremely or the production of other macromolecules, which are required for tumor proliferation. The discovery of oncometabolites completed with the discovery of cancer metabolic biomarkers, which is found in salvia, plasma, urine, and tissue specimens. Some of these cancer biomarkers are glutamate, isoleucine, threonine, histidine, and mannose in breast cancer and lactate, isoglutamine, taurine, tyrosine, triglycerides, and lipids in colorectal cancer. Indeed, metabolic biomarkers help scientists to discover novel methods of cancer management and fundamental information about tumor staging [48].

7 Cancer Cells and OMICs Revolution

OMICs technologies are changing medicine and giving insight into the uncovering of innovative diagnosis, prognostic and therapeutic indicators that will enhance patient outcomes, treatment of cancer and rare diseases. Cancer treatment and clinical medicine transformed by OMICs technology describing the use of these technologies in proteomics (analyzing proteins), genomics (analyzing the entire genes), epigenomics (analyzing the epigenetic modifications), pharmacogenomics (analyzing the role of genome in drug response), transcriptomics (analyzing RNA molecules transcripts), metabolomics (analyzing the chemical processes involving metabolites), lipidomics (analyzing the whole lipids in cells), and so on. The data of cancer OMICs are rapidly collected and provided a valuable resource for identifying novel targets [49]. The OMICs technology accumulated data from each bimolecular class including nucleic acids (DNA and RNA), proteins, and lipids. In addition, genomics profiling technologies have the ability to translate into the clinic and can analyze different relative cancer genes in parallel and reduce important factors such as time, costs, and equipment which are needed in conventional testing. The main aim of the project of sequencing the whole genomes (100,000 genome project) of patients with collecting high quality samples for future "OMICs" research is adding the depth to the clinical interpretation aspects of cancer and rare diseases [50]. Scientists hope that with future analyzing of OMICs data, we can increase our therapeutic productivity for molecular targets of cancer therapies. Each of these molecular classes focuses on their own field in OMICs study. The information gained from these classes help in increasing the knowledge of diseases and more importantly cancer. It must be admitted that there are several ways in cancer medicine which apply molecular-based assays. Recent developments in the RNA analysis and mutation testing suggest new cancer detection tools of unknown primary site. Examples of molecular testing currently relevant to clinical practice or research in oncology are indicated in Table 2. Although the availability of these

Table 2 Molecular testing in cancer research [51–53]

Circulating tumor fragments	Example of markers	Molecular tests	Effect in cancer
DNA	MYC∗-HER2 PTEN∗-APC∗	PCR∗, ARMS∗, NGS∗, CGH∗, FISH∗	Amplification Loss of heterozygosity
RNA	TP53∗, KRAS, BRCA2∗, AKT∗, CDKN1A∗, P13K∗, luminal A, versus triple-negative breast cancer	PCR, NGS, microarray	Induction, repression, missense/nonsense
Protein	ERK∗/*P*Rb∗	Protein (IHC∗) assays, proteomic profiles	Upregulation/downregulation
Lipid	Phosphatidylcholine/phosphatidylethanolamine	Mass spectrometry NMR-based sensors Optical biosensors Electrochemical biosensors Mechanical biosensors	Increase/decrease

MYC∗ A family of regulator genes and proto-oncogenes that code for transcription factors, *PTEN*∗ Phosphatase and tensin homolog, *APC*∗ Adenomatous polyposis coli, *TP53*∗ Tumor protein 53, *BRCA2*∗ Breast cancer susceptibility gene 2, *AKT*∗ Protein kinase B, *CDKN1A*∗ Cyclin-dependent kinase inhibitor 1, *PI3Ks*∗ Phosphoinositide 3-kinases, *ERK*∗ Extracellular signal regulated kinase, *PRb*∗ Retinoblastoma protein, *PCR*∗ Polymerase chain reaction, *ARMS*∗ Amplification-refractory mutation system, *NGS*∗ Next generation sequencing, *CGH*∗ Comparative genomic hybridization, *FISH*∗ Fluorescence in situ hybridization, *IHC*∗ ImmunoHistoChemistry information exceeds our ability to apply it therapeutically, recent advancement in OMICs technologies has provided some opportunities to unveil the secrets of cancer genome and make a great impact on treatment. Despite the fact that clinical applications are limited currently, in the next decade we will see significant progress based on better understanding of OMICs technology [51, 52].

8 Diagnosis and Treatment of Cancer Using OMICs Technology

Currently, a surge of interest is associated with the potential contribution of OMICs technologies in cancer diagnosis. In fact, cancer is a complex disease whose different factors, such as genetics, cellular, and environmental effects, play a key role in its development. Recently, advancements in new technologies such as genomics, epigenomics, metabolomics, and bioinformatics provided new insights in cancer diagnosis and cancer therapy. Indeed, now the treatment teams are potent to achieve more detailed and accurate results of both quantitative and qualitative molecular evaluation due to the new analysis tests. As a result, these researches enhance our data about the function of disease in patient's body and help us with making the best accurate decision for choosing the most effective treatment method and clinical care pathways [6, 50, 54].

9 Developing Novel Treatment for Cancers

According to the cancer statistics presented worldwide, in the more developed countries there has been a dramatic growth in the rates of patients suffering from cancer. Although researchers and clinicians have discovered and analyzed many therapeutic strategies to cure cancer, more effective methods are required. The most common cancer therapies are surgery, chemotherapy, and radiation therapy. Scientists believe that some side effects such as toxicity and drug resistance are inevitable during chemotherapy and these two side effects are the most expected problems [55]. The traditional methods in cancer treatments mostly involve chemotherapy, with the main aim to maximize damage to the rapidly dividing and growing cancerous cells and tumors. Although these methods often extend the life of patient, the normal cells are damaged during treatment which will show the adverse effects in various tissues and organs. These treatment approaches can be prognosed, diagnosed, or used in cancer treatment. The FDA has been permitted the derived products from these approaches for use in clinical treatments. However, these treatments induce some critical side effects such as cardiac disorders, thrombosis, and hypertension. Therefore, we need some of the new cancer treatment modalities. New methods such as OMICs technologies increase the overall survival and improve the cancer management in patients suffer from cancer and tolerated better than traditional chemotherapy. Today, by the advancements in tumor biology, discovering new anticancer drugs and novel technologies, scientists are able to target cancer cells. The main purpose of the new treatments is to impact cancer cells and their function while keeping normal cells, with lowest side effects. Finding the phenotypic properties of tumors, their signaling pathways, their vascular system, and the destructive effect on immune system help scientists to find new drug targets. Novel treatment methods such as inhibition of proliferative signaling, immunotherapy, targeting therapy, enzymatic inhibitory drugs, targeting chromosomal aberrations are being tested both in research studies and in clinical trials. In all, in the management of cancer and following symptoms, combination of old methods and novel therapeutic methods can be used. Treatment of cancer will accelerate with developed diagnostic technologies and advanced novel methods in near future [56, 57].

10 Conclusion and Future Direction

Cancer is a complex disease that caused by accumulation of several factors such as genetics, epigenetics, proteomics, and transcriptional alterations, which result in metabolic and cellular damage [3, 4, 50].

Across the world, the impact of cancer on health society is undeniable based on statistics. On the other hand, the number of people who survived from cancer due to early diagnosis, monitoring, and proper treatment is increasing. These findings are achieved using new techniques in all steps of the treatment process. As the diagnosis and treatment of cancer are timely, costly, and crucial, early diagnosis importantly plays role in treatment or improving outcomes of cancer [1].

Genetics, proteomics, and metabolomics profiling could be considered using OMICs technologies in both early diagnosis and monitoring the patients suffering from cancer. These techniques are able to find the specific biomarkers of any disease or cancer. Detection of circulating biomarkers helps the treatment team to find the stage of the disease and the way of treatment [7, 34].

Genomics is a field of OMICS focusing on the structure, function, evolution, mapping, and editing of genomes. In addition, using DNA sequencing and bioinformatics, we can apply sequencing and analysis of entire genomes. In final step, genomics helps specialists to accumulate and analyze the function and structure of entire genomes. This information in genomics increases our knowledge about how tumors evolve during progression and under the pressure of cancer therapy [36].

Among the mentioned techniques, proteomics identifies the protein changes caused by disease process and analyze the entire set of proteins. Therefore, developing proteomics technology facilitates analyzing the full set of proteins expressed by tumor cells [41].

Since cancer is a metabolic disorder, cellular metabolic pathways and metabolites are changed during the disease process. In this respect, metabolomics is the study of metabolites, which are small molecules found in tissues and bio-fluids of an organism. Accordingly, with the advancement of cancer detection, metabolomics can be used to define metabolic alteration during carcinogenesis and find potential treatments in early stages diagnosis of cancer. Therefore, scientists have found deeper understanding of cancer metabolites via metabolomics [44, 45].

Currently, clinical medicine and cancer treatment have been transformed by OMICs technology. Scientists hope that in future, by analyzing of OMICs data, new therapeutic methods will be found for molecular targets in cancer. The obtained data from various molecular classes by novel methods exceeds our ability to apply them therapeutically [55–57].

Acknowledgement The authors would like to acknowledge Maryam Afshari and Dr. Mohsen Khorshidi for their kind support.

References

1. Swann R, et al. Diagnosing cancer in primary care: results from the National Cancer Diagnosis Audit. Br J Gen Pract. 2018;68(666):e63–72.
2. Siegel R, et al. Cancer statistics, 2019. CA Cancer J Clin. 2019;69(1):7–34.
3. Miller K, et al. Cancer treatment and survivorship statistics. CA Cancer J Clin. 2016;66(4):271–89.
4. Karczewski K, Snyder M. Integrative omics for health and disease. Nat Rev Genet. 2018;19(5):229.
5. Yoo B, et al. Clinical multi-omics strategies for the effective cancer management. J Proteome. 2018;30(188):97–106.
6. Cho WC-S, editor. An omics perspective on cancer research. Dordrecht: Springer; 2010. p. 1–9.
7. Rapisuwon S, Vietsch E, Wellstein A. Circulating biomarkers to monitor cancer progression and treatment. Comput Struct Biotechnol J. 2016;1(14):211–22.
8. Yu K, Snyder M. Omics profiling in precision oncology. Mol Cell Proteomics. 2016;15(8):2525–36.
9. Holzinger A, Dehmer M, Murisica I. Knowledge discovery and interactive data mining in bioinformatics-state-of-the-art, future challenges and research directions. BMC Bioinformatics. 2014;15(6):1.
10. Moreno-Sánchez R, et al. Understanding the cancer cell phenotype beyond the limitations of current omics analyses. FEBS J. 2016;283(1):54–73.
11. Zhang X, et al. Mass spectrometry-based "omics" technologies in cancer diagnostics. Mass Spectrom Rev. 2007;26(3):403–31.
12. Radpour R, Forouharkhou F. Single-cell analysis of tumors: creating new value for molecular biomarker discovery of cancer stem cells and tumor-infiltrating immune cells. World J Stem Cells. 2018;10(11):160.
13. Carlomagno N, et al. Diagnostic, predictive, prognostic, and therapeutic molecular biomarkers in third millennium: a breakthrough in gastric cancer. Biomed Res Int. 2017;2017:7869802.
14. Wang D, Bodovitz S. Single cell analysis: the new frontier in 'omics'. Trends Biotechnol. 2010;28(6):281–90.
15. Milestones in cancer research and discovery. 2015 Jan 21. https://www.cancer.gov/research/progress/250-years-milestones.
16. Saijo N, et al. New strategies for cancer therapy in the 21st century. Cancer Chemother Pharmacol. 2001;48(1):102–6.
17. Kangwan N, et al. Chemoquiescence for ideal cancer treatment and prevention: where are we now? J Cancer Prev. 2014;19(2):89–96.

18. Charmsaz S, et al. Innovative technologies changing cancer treatment. Cancers (Basel). 2018;10(6):208.
19. The American Cancer Society Medical and Editorial Content Team. The history of cancer; evolution of cancer treatments: surgery. 2014 June 12. https://www.cancer.org/cancer/cancer-basics/history-of-cancer/cancer-treatment-surgery.html.
20. Stanford Health Care. Types of surgery for cancer treatment. https://stanfordhealthcare.org/medical-treatments/c/cancer-surgery/types.html.
21. Subotic S, Wyler S, Bachmann A. Surgical treatment of localised renal cancer. Eur Urol Suppl. 2012;11(3):60–5.
22. Baskar R, et al. Cancer and radiation therapy: current advances and future directions. Int J Med Sci. 2012;9(3):193–9.
23. Chiang T, et al. Complete remission in very advanced oral cancer by docetaxel, cisplatin, 5-fluorouracil based induction chemotherapy followed by concurrent chemoradiation. J Dent Sci. 2018;13(1):82–4.
24. Yang Y. Cancer immunotherapy: harnessing the immune system to battle cancer. J Clin Invest. 2015;125(9):3335–7.
25. Armitage EG, Southam AD. Monitoring cancer prognosis, diagnosis and treatment efficacy using metabolomics and lipidomics. Metabolomics. 2016;12(9):146.
26. Triulzi T, Bianchi GV, Tagliabue E. Predictive biomarkers in the treatment of HER2-positive breast cancer: an ongoing challenge. Future Oncol. 2016;12(11):1413–28.
27. Wang Y, et al. UGT1A1 predicts outcome in colorectal cancer treated with irinotecan and fluorouracil. World J Gastroenterol. 2012;18(45):6635–44.
28. Abrahao-Machado LF, Scapulatempo-Neto C. HER2 testing in gastric cancer: an update. World J Gastroenterol. 2016;22(19):4619–25.
29. Prevodnik VK, et al. The predictive significance of CD20 expression in B-cell lymphomas. Diagn Pathol. 2011;6:33.
30. Zheng W, et al. CD30 expression in high-risk acute myeloid leukemia and myelodysplastic syndromes. Clin Lymphoma Myeloma Leuk. 2013;13(3):307–14.
31. Minamimoto R, et al. Prediction of EGFR and KRAS mutation in non-small cell lung cancer using quantitative 18F FDG-PET/CT metrics. Oncotarget. 2017;8(32):52792–801.
32. Bhatia P, et al. Impact of BRAF mutation status in the prognosis of cutaneous melanoma: an area of ongoing research. Ann Transl Med. 2015;3(2):24.
33. Wu L, Qu X. Cancer biomarker detection: recent achievements and challenges. Chem Soc Rev. 2015;44(10):2963–97.
34. Sani M-RM, et al. Biomarkers in cancer. Immunopharmacogenetics. 2018;1:01.
35. Ludwig J, Weinstein J. Biomarkers in cancer staging, prognosis and treatment selection. Nat Rev Cancer. 2005;5(11):845.
36. Wills Q, Mead AJ. Application of single-cell genomics in cancer: promise and challenges. Hum Mol Genet. 2015;24(R1):R74–84.
37. Romanov V, et al. A critical comparison of protein microarray fabrication technologies. Analyst. 2014;139(6):1303–26.
38. Barbulovic-Nad I, et al. Bio-microarray fabrication techniques--a review. Crit Rev Biotechnol. 2006;26(4):237–59.
39. Russell S, Meadows L, Russell R. Microarray technology in practice. Cambridge: Academic; 2008.
40. Wang Z, Gerstein M, syder M. RNA-Seq: a revolutionary tool for transcriptomics. Nat Rev Genet. 2009;10(1):57.
41. Srinivas P, et al. Proteomics for cancer biomarker discovery. Clin Chem. 2002;48(8):1160–9.
42. Ostroff RM, et al. Unlocking biomarker discovery: large scale application of aptamer proteomic technology for early detection of lung cancer. PLoS One. 2010;5(12):e15003.
43. Auburn RP, et al. Robotic spotting of cDNA and oligonucleotide microarrays. Trends Biotechnol. 2005;23(7):374–9.
44. Mehrparavar B, et al. Metabolomics of male infertility: a new tool for diagnostic tests. J Reprod Infertil. 2019;20(2):64–9.

45. Agharezaee N, et al. Metabolomics: a bird's eye view of infertile men. Tehran Univ Med J TUMS Publ. 2018;75(12):860–8.
46. Khatami F, et al. Oncometabolites as biomarkers in thyroid cancer: a systematic review. Cancer Manag Res. 2019;11:1829.
47. Hammoudi N, et al. Metabolic alterations in cancer cells and therapeutic implications. Chin J Cancer. 2011;30(8):508.
48. Wishart D, et al. Cancer metabolomics and the human metabolome database. Meta. 2016;6(1):10.
49. Vucic EA, et al. Translating cancer 'omics' to improved outcomes. Genome Res. 2012;22(2):188–95.
50. Yoo BC, et al. Clinical multi-omics strategies for the effective cancer management. J Proteome. 2018;188:97–106.
51. Epstein RJ, Lin F. Cancer and the omics revolution. Aust Fam Physician. 2017;46(4):189–93.
52. Sokolenko AP, Imyanitov EN. Molecular diagnostics in clinical oncology. Front Mol Biosci. 2018;5:76.
53. Kubicek-Sutherland JZ, et al. Detection of lipid and Amphiphilic biomarkers for disease diagnostics. Biosensors (Basel). 2017;7(3):25.
54. Karczewski KJ, Snyder MP. Integrative omics for health and disease. Nat Rev Genet. 2018;19(5):299.
55. M Toloudi PA, Chatziioannou M, Kourtidou E, Vlachou I, Mimikakou G, Chlichlia A, Papasotiriou I. Recent developments in cancer treatment: a review. Pharmaceut Reg Aff. 2014;S1:001.
56. Ramaswami R, Harding V, Newsom-Davis T. Novel cancer therapies: treatments driven by tumour biology. Postgrad Med J. 2013;89:652–8.
57. Chakraborty C, et al. The novel strategies for next-generation cancer treatment: miRNA combined with chemotherapeutic agents for the treatment of cancer. Oncotarget. 2018;9(11):10164–74.

Genomics, Proteomics, and Metabolomics of Cancer Stem Cells (CSCs)

Fatemeh Khatami, Seyed Mohammad Tavangar, and Navaz Karimian Pour

Abstract Cancer stem cells (CSCs) are just projected as the cancer triggering cells in charge of both tumor genesis and cancer resistance. While the theory of CSCs originates from that of normal stem cells, CSC is not necessarily aberrant counterparts of normal stem cells. CSCs can be the origin of circulating tumor cells (CTCs) which are releasing from tumor and shed into the vasculature or lymphatic. CTC and CSC are suggested as tools for recognition and classification of disease and individualization of therapy in patients with many solid tumors. In fact, the genetic and epigenetic of CSCs are different from both normal stem cells and tumoral cells. In order to find the best discriminating differences and basic tumor genesis pathway we should know the comprehensive profile of genomics, proteomics, metabolomics, transcriptomics, and epigenomics. Over the last decade, key advancements in omic have assisted high-throughput monitoring of a diversity of molecular and organismal processes. To date, a variety of software have been developed to make effective integration of OMICS-based analyses of several solid tumors to the new targeted cancer therapies. In this chapter we have discussed the CSCs, CTCs from OMICS perspectives.

Keywords Cancer Stem cells · Omics · Genetic · Epigenetic

F. Khatami (✉)
Chronic Diseases Research Center, Endocrinology and Metabolism
Population Sciences Institute, Tehran University of Medical Sciences, Tehran, Iran
e-mail: f-khatami@alumnus.tums.ac.ir

S. M. Tavangar
Department of Pathology, Dr. Shariati Hospital,
Tehran University of Medical Sciences, Tehran, Iran
e-mail: tavangar@ams.ac.ir

N. Karimian Pour
Leslie Dan Faculty of Pharmacy, University of Toronto, Toronto, ON, Canada
e-mail: Navaz.karimianpour@mail.utoronto.ca

B. Arjmand (ed.), *Genomics, Proteomics, and Metabolomics*, Stem Cell Biology and Regenerative Medicine, https://doi.org/10.1007/978-3-030-27727-7_9

1 Introduction

The knowledge of tumor is changed because of the new era of "omics"—genomics, epigenomics, proteomics, metabolomics, and lipidomics are considered more than before. Over the last years, many efforts have been done to reveal the molecular mechanisms of carcinogenesis involving single OMICS approaches like searching the genome for cancer-specific genetic and epigenetics alterations through altered epigenetic-landscapes within cancer cells or by discovering the discriminative expression of mRNA and protein over transcriptomics and proteomics techniques. Although the single-level omics methods are involved in identification of cancer-specific genetic mutations and epigenetic modifications they do not have resolving-power to find the logical connection between molecular signatures and the phenotypic indicator of cancer hallmarks. Whereas the multi-omics approaches concerning the examination of the cancer cells/tissues in several scopes can uncover the complicated molecular mechanism underlying different phenotypic manifestations of cancer hallmarks such as metastasis and angiogenesis. There is a big hope that in the near future, systems biology based approach linked to omics can predict the changes of cancer cells after chemo-/immunotherapy treatment.

2 Cancer Stem Cells (CSCs)

Normally, stem cells are undifferentiated cells with a significant potential to progress into different cell types in the body during early life and growth. In several tissues of normal alive animal and human they work as a class of internal repair system, dividing fundamentally without limit to replenish other cells. Once a stem cell divides, each new cell can decide whether it can remain as a stem cell or become another type of differentiated cell with a more specific function, such as a muscle cell or red blood cell [1]. The most important and useful characteristic of stem cells is their self-renewal property which is the basic of tumor formation. In fact, remarkable parallels can be found between stem cells and cancer cells: tumors may often originate from the transformation of normal stem cells, similar signaling pathways may control self-renewal in stem cells and cancer cells, and cancer cells include "cancer stem cells or CSCs" as the rare cells with indefinite potential for self-renewal that drive tumorigenesis [2]. It is hypothesized that subsequently of asymmetric division, daughter tumor cells undertake some degree of differentiation (Fig. 1). Numerous pathways are linked to tumorigenesis which are similar to the stem cell self-renewal and differentiation process [3]. By way of illustration, mutations and alterations in the Wnt, TGF-β/BMP, notch, and integrin signaling pathways have been known in some tumor systems including hematopoietic, gastrointestinal, breast, and prostate cancers [4]. These pathways are also concerned in stem cell self-renewal system and differentiation in numerous systems.

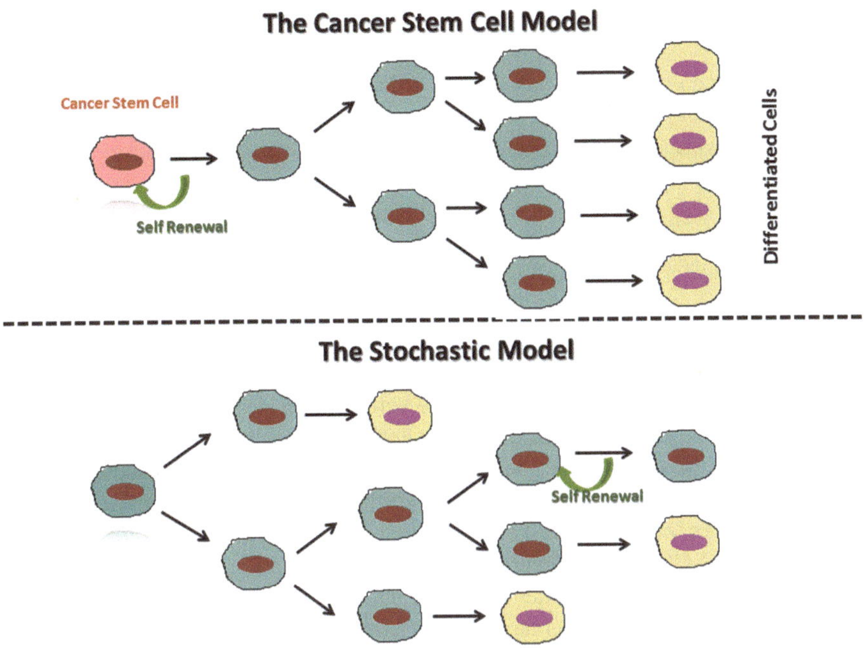

Fig. 1 Two models are suggested for heterogeneous potential of tumor cells. The CSCs in the stochastic model, mutations accumulate over time and some cell can have tumorigenic potential

The similar genomics, proteomics, and metabolomics profile of normal stem cells and tumor cells is headed to the suggestion that normal stem cells are the target of mutagenesis leading to tumor formation.

3 Circulating Tumor Cells (CTCs)

Circulating tumor cells (CTCs) are the group of tumor cells that originated from primary solid tumors and shedding into blood stream [5]. Metastases are main cause of cancer-related deaths and CTCs can seeds the subsequent growth of additional tumors in distant organs [6]. For the first time CTCs were reported in 1869 in the blood of a man with metastatic cancer by Thomas Ashworth "cells identical with those of the cancer itself being seen in the blood may tend to throw some light upon the mode of origin of multiple tumors existing in the same person" [7]. However, the importance of CTCs in modern cancer research and management is started from 1990 with the demonstration of early presence of CTCs in primary stage of tumor formation [8]. Nowadays in modern cancer research CTCs are described as the main liquid biopsy component and understanding the CTCs biological properties, genomics, proteomics, and metabolomics has revealed the serious part of tumor metastatic [9, 10].

3.1 Functional Characteristics of Cancer Stem Cells

For the first time in 1994, through a study of human acute myeloid leukemia it was characterized that stem cells have role in tumor genesis [11]. After that in 2003, human CSCs were identified in solid tumors like breast and brain tumors and then in a wider cancer types, including colon, pancreas, lung, prostate, melanoma, and glioblastoma [12, 13]. It was shown by Muhammad Al-Hajj that just 100 cells with phenotype $CD44^{+}CD24^{-/low}$ $Lineage^{-1}$ were able to form tumors in mice, while more than 10,000 cells failed to form tumors in non-obese diabetic/severe combined immune-deficient (NOD/SCID) mice [12]. The cell surface expressed proteins were principal of CSCs isolation (CD44, CD24, CD29, CD90, CD133, epithelial-specific antigen (ESA), and aldehyde dehydrogenase1 (ALDH1)) [14]. CSCs are highlighted by their capacity to generate more self-renewal property and asymmetric cell division achieves both tasks, as one progeny retains with self-renewal property identity, while the other undergoes rounds of cell division and subsequent post-mitotic differentiation. At first it was believed that CSCs represent a minor portion of the total tumor, but it has been appealed that one-fourth of cancer cells can have the properties of CSCs [15]. Several theories are presented for the CSCs origin. As we mentioned above the first one believes that CSCs rise from normal stem cells in which some genetic mutations happen and earn the power to create tumors [16]. Some CSCs exhibit similarities to normal stem/progenitor cells in cellular property, phenotype, function, and even cell surface markers. For example, the $CD44^{+}/CD24^{-/low}$ cell population identified as mammary gland progenitor cells resembles the $CD44^{+}CD24^{-/low}$ $Lineage^{-}$ cells used to identify CSCs from breast cancer patients. The second theory for the origin of CSCs opposing recommends that CSCs rise from normal somatic cells which obtain stem-like features and malignant behavior over genetic and epigenetic alterations. For supporting this theory the epithelial–mesenchymal transition (EMT) can be presented. The EMT is an extremely conserved mechanism that converts epithelial cells into mesenchymal cells critical for normal embryogenesis and tissue repair and tumor progression as well [17]. EMT is thoroughly connected to the cancer recurrence and metastasis and suppression of nestin, fibroblast growth factor (FGFR)-2, FGFR-2 IIIc, and regulation of the EMT using epithelial splicing regulatory protein 1 (ESRP1) are operative in the treatment of immunodeficient mice with pancreatic cancer [18]. Understanding the biology of CSCs and treatment resistance or tumor recurrence is completely dependent on CSCs molecular profiling and CSCs omics, including genomics, epigenomics, methylomics, proteomics, or metabolomics. The words ending with -ome are used for items of study of the genome, proteome, transcriptome, or metabolome [19].

The new breakthrough entitled "trans-omic" analysis with the aim of restructuring overall biochemical networks through several omic layers by use of both

[1] *$CD44^{+}CD24^{-/low}$ $Lineage^{-}$*: A subpopulation of breast cancer isolated based on flow cytometry and cell surface antigens.

multi-omic measurements and computational data integration, and several trans-ome-wide association study (trans-OWAS) can connect the phenotypes to the trans-omic that replicate both genetic and environmental factors [20].

4 Cancer Stem Cells Genomics

Genomics is the science that studies the structure, function, evolution, and mapping of genomes and aims at characterization and quantification of genes, which direct the production of proteins with the assistance of enzymes and messenger molecules. In fact the combination of recombinant DNA, DNA sequencing methods (next generation sequencing or NGS) together with bioinformatics is used for the genomes analysis. Opposing to "classical genetics" that studies one gene or one gene product at a time, genomics highlighted the interactions between loci and alleles within the genome [21]. The Cancer Genome Atlas (TCGA) is working on cancer genome as collaboration between the National Cancer Institute (NCI) and the National Human Genome Research Institute (NHGRI). TCGA makes comprehensive, multidimensional maps of the key genomic changes in 33 types of cancer. The TCGA dataset, comprising more than two petabytes of genomic data, has been made publically available, and this genomic information helps the cancer research community to improve the prevention, diagnosis, and treatment of cancer [22]. TCGA dataset was stated as a pilot project in 2006, its instruction was to make a complete landscape of genetic and epigenetic changes in all tumor types. After 10 years and a total of $375 million investment, TCGA has incorporated scientific contributions from more than 150 researchers from 16 countries, characterizing 10,000 tumors from more than 25 different cancer types. Its 20 petabytes of data include ten million mutations, and they have been reported in 17 (so far) publications from the TCGA Research Network and mined in hundreds of articles [23]. Moreover the cBio Cancer Genomics Portal (http://cbioportal.org) is an open-access resource for collaborative search of multidimensional cancer genomics datasets, now it is offering access to data of 5000 tumor samples from 20 cancer researches. Actually, the cBio Cancer Genomics Portal meaningfully decreases the barriers between complex genomic data and cancer researchers who want rapid, intuitive, and high-quality access to molecular profiles and clinical attributes from large-scale cancer genomics projects and authorizes researchers to translate these rich datasets into biologic visions and clinical applications [24]. There are some candidate genes as the key regulating element in both stem cells and cancer development.

Some studies published thatsome data about the cancer genome-wide expression profile. There are two different datasets in which the genome-wide expression data of $CD133^+$ cells from human glioblastomas are presented [25]. Both research offered around 100 candidate genes, which were differentially expressed in the $CD133^+$, while just ten genes are common in both studies (Fig. 2) [26, 27]. Available Search Tool for the Retrieval of Interacting Genes/Proteins (STRING) offers indication that nine out of these ten candidate genes form an interactive network

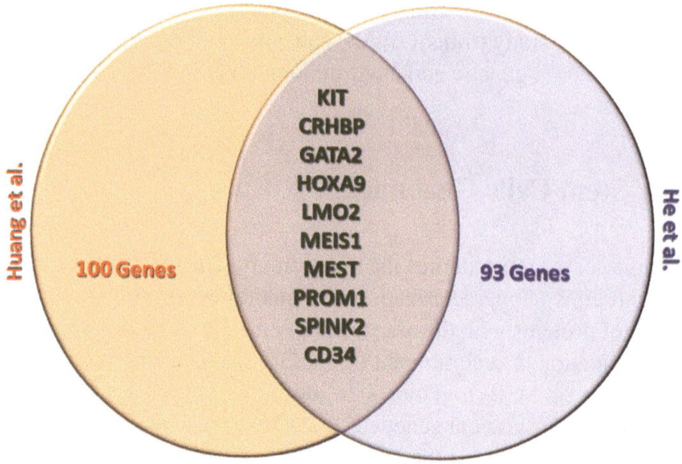

Fig. 2 Ten mutual genes of two different genome-wide expression data studies by He, and by Huang

straight network with each other. These target genes mainly belong to proliferative pathways (JAK/STAT[2]) and some pathways involving in cell differentiation (HOX,[3] PBX,[4] MEIS,[5] GATA2[6]).

One remarkable receptor is the proto-oncogene *C-KIT* which its mutation is linked to the outcome and progression in gliomas and acute myelogenous leukemia [28–30]. Contrary to *MEST*[7] that is not associated with the proposed network of the other nine genes. Imprinting of MEST is observed in normal cells, and a loss of imprinting of MEST has resulted in several tumorgenesis pathways [31, 32]. Some additional respective datasets by DAVID[8] and KEGG[9] represented some significant enhanced scores of several pathways [33, 34]. The highlighted pathways include the glycolysis/gluconeogenesis (HSA00010), cell communication (HSA01430), proteasome (HSA03050), cell-cycle (HSA04110), Wnt-signaling (HSA04310), TGF-β-signaling (HSA04350), and apoptosis (HSA04210) [25].

There are common signaling pathways, e.g., Wnt, NOTCH, SHH, JAK/STAT, in link with several CSCs [35, 36]. An integrative analysis of multi-marker genomics

[2] *JAK-STAT signaling pathway*: Janus kinases (JAKs), signal transducer and activator of transcription (STAT) proteins.

[3] *Hox genes*: A subset of homeotic genes are a group of related genes that control the body plan of an embryo along the head–tail axis.

[4] *PBX*: Pre-B cell leukemia transcription factor.

[5] *MEIS*: Myeloid ecotropic viral integration site 1 homolog.

[6] *GATA2*: GATA binding protein 2.

[7] *MEST*: Mesoderm specific transcript homolog.

[8] *DAVID*: The database for annotation, visualization, and integrated discovery.

[9] *KEGG*: Kyoto Encyclopedia of Genes and Genomes.

data over the ovarian cancer stem cell (OCSCs) indicated to the 22 activated transcription factors and 15 of them are described in association with CSCs [37]. The projected approach can reveal the activated upstream signaling, activated transcription factors that are critical signaling pathways of ovarian CSC [38].

It is shown that the ovarian cancer cells transduced with the four Yamanaka factors[10] the pluripotent stem cell properties and drug resistance increased. The CSCs with overexpressed Yamanaka factors are called induced ovarian CSCs (iOCSCs) and have the ability to comprehend up to at least 40 passages [39].

A genome editing approach by CRISPR/Cas9 technology in human colorectal cancer (CRC) organoids[11] that carry EGFP and lineage-tracing cassettes knocked in the LGR5 locus indicted to similar gene expression pattern of normal intestinal stem cells that they propagate the disease to recipient mice powerfully [40].

Because of the CTCs importance in cancer managements recently several attempts were done through whole exome sequencing for the identification of genomic signatures [41, 42]. In metastatic castration-resistant prostate cancer (mCRPC) genomic gains in more than one-fourth of CTCs were observed in AR, FOXA1, ABL1, MET, ERG, CDK12, BRD4, and ZFHX3, whereas genomic losses involved PTEN, ZFHX3, PDE4DIP, RAF1, and GATA2 [43, 44]. There are some new technological breakthroughs in cell capture and single-cell-sequencing (scDNA-seq) protocols that make it probable to interrogate the information of genome, transcriptome, and epigenome of CTCs [45]. scDNA-seq of CTCs and mutation detection make it possible to recognize differences of primary and metastatic tumors [46–48]. Most published studies estimated the technical inconsistency relating the variants obtained from single-cell data with bulk sequencing or control samples [41]. Single nucleotide polymorphisms (SNPs) can be distinguished exactly from (scDNA-seq) that are based on the mutation frequency across cells to calculate the posterior probability of the variant to be present in at least two cells [49–51]. The DNA copy number variations (CNVs) method is based on segmentation algorithm and GC-normalized coverage [52–55].

5 Cancer Stem Cells Proteomics

Proteomics is the proteome-based method to study the proteins in a large-scale [56, 57]. In comparison with genomics the exact word of "proteomics" was created in 1997, and the first dedicated proteomics laboratory in 1995 was created at Macquarie University [58, 59]. Proteomics combines progressive analytical technology with novel approaches of calculation and can precisely recognize and enumerate

[10] *Yamanaka factors*: Oct3/4, Sox2, Klf4, c-Myc as the highly expressed protein in embryonic stem (ES) cells inducing pluripotency in both mouse and human somatic cells and regulating the developmental signaling network necessary for ES cell pluripotency.

[11] *Organoid*: Tiny, self-organized three-dimensional tissue cultures that are derived from stem cells.

thousands of proteins in several biological samples [60, 61]. There are two common primary techniques of proteomics: protein chip-based proteomics and mass spectrometry (MS)-based proteomics. Protein chip-based proteomics which is also called microarray-based proteomics is done based on molecular interaction on the surface of solid carriers [62, 63]. In spite of the fact that it is fast and high-throughput, technique with good miniaturization, the problem of non-characteristic combinations and false-positive detection still remains [64–66]. Therefore, MS-based proteomics technology, including label-based and label-free approaches, has been the typical technology for large-scale protein documentation and quantitation. Many label-based proteomics approaches are available now like isotope-coded affinity tags (ICAT), isobaric tags for relative and absolute quantification (iTRAQ), tandem mass tags (TMT), and stable isotope labeling with amino acids in cell culture (SILAC) [67].

In CSCs proteomics can improve knowledge of deregulated or unregulated pathways as a result of changed protein expression profile. The study of CSCs in breast cancer in comparison with non-CSC in the same tumor highlighted the Ferritin heavy chain1 (FTH1) depletion meaningfully reduced the self-renewal of CSCs [68]. During a proteomic analysis of CSCs fractions and the mature luminal cells (CD49f-EpCAM^{+}) from the MCF7 cell line, a total of 3304 proteins were highlighted which mostly were involved in tumor genesis, cancer migration, energy metabolism, cancer invasion, and cancer progression [69]. Ten annexin proteins were identified in which five of them presented noteworthy differences and the differential expression of ANXA3 was confirmed by Western blot analysis and IHC [69]. In the prostate cancer cell line DU145 which was isolated into CD44^{+} or CD44^{-} cells two proteins cofilin and annexin A5 were shown in association with proliferation and metastasis [70]. A recent comparative proteomics of CSCs in osteosarcoma by ultra-high-performance liquid chromatography (HPLC) and Orbitrap Fusion mass spectrometer specified the differential α-actinin 4 (ACTN4) proteins as the candidate one [71]. Differential proteomic analysis of CSCs in two clones of Hep3B cell lines by isobaric tag labeling and mass spectrometry resulted in expressions of S100P, S100A14, and vimentin [72]. Proteomics analysis of gastric cancer stem cells the eight proteins, RBBP6, GLG1, VPS13A, DCTPP1, HSPA9, HSPA4, ALDOA, and KRT18 of which RBBP6 was suggested to be a promising prognostic biomarker and a therapeutic target for gastric cancer [73]. In colon cancer 3048 proteins were identified as the different ones between colon cancer stem cells (CCSCs) and differentiated tumor cells (DTCs) in which 32 proteins were at least two fold upregulated [74]. Additional pathway analysis presented that "cell death" regulation is unusually different between the two targeted cell types and the top upregulated proteins were BIRC6 (inhibitor of apoptosis proteins). So, knockdown of BIRC6 sensitized CCSCs against the chemotherapeutic drugs oxaliplatin and cisplatin [74].

There are some complications for CTCs identification, isolation, and characterization and it is really essential to discuss the protein expression profile and some of the highlighted protein markers. There are some recent protein markers that are related to the overall population of the CTCs including EpCAM, N-cadherin, vimentin, integrin αVβ6, β-catenin, SOX10, snail family, and twist 1 [75]. EpCAM (CD326) is a type I transmembrane glycoprotein expressed on the basolateral surfaces of epithelial cells and having role in homophilic Ca2+-independent cell adhesion molecule. The proteomic effects of specific cancer-related mutations have been categorized by Halvey et al. and the proteomic consequences of a single gene mutation using an isogenic colorectal cancer cell culture model were discussed [76]. It was additionally described that SW480Null cells had decreased levels of many other proteins involved in adherents junction formation and cell adhesion, including CDH3, CTTNA1, CTTND1, NCAM1, EVPL, DSG2, CECAM1, and, above all, EpCAM. The colon CSCs isolated and spread in serum-free stem cell culture conditions took the expression of well-known cell surface markers, including CD133, CD166, CD44, and EpCAM, and additional stem cell associated proteins, such as NES, BMI-1, and MSI-1 [77]. Using an extensive set of LC/MS/MS (the combination of liquid chromatography (LC) with mass spectrometry (MS)) analyses to maximize glycoprotein identification and demonstrated that the glycoprotein profiles of several human breast cancer cell lines are dissimilar from normal/benign breast cell lines are dissimilar from normal/benign breast cell lines [78]. Interestingly, EpCAM was present in all breast cancer cell lines but not in normal human mammary epithelial cells (HMECs). In fact N-cadherin (cadherin-2/CD325/cluster of differentiation 325) is a transmembrane protein expressed in multiple tissues and functions to mediate cell–cell adhesion and is protein of CTCs and CSCs and acts as a fundamental molecule during the EMT. It was shown that lowering the expression of E-cadherin in the gut epithelium results in decreased proliferation and increased apoptosis of epithelial cells so it can be the important factor of tumor genesis [79, 80]. It was shown that the expression of E-cadherin, decreased in the gut epithelium have result in increased apoptosis of epithelial cells, so it can be the important factor of tumor genesis [81].

Vimentin is a structural protein encoded by the VIM gene, it is a type III intermediate filament (IF) protein that is usually expressed in mesenchymal cells and together with tubulin-based microtubules and actin-based microfilaments involves the cytoskeleton. Normally, vimentin has role in immune system and is secreted into the extracellular space by macrophages in response to inflammatory cytokines [82]. Additional MS (mass spectrometry) analysis information suggested the ligand for the NKp46 receptor of natural killer cells [83, 84]. Moreover, vimentin is newly stated for direct interaction with NKp46 in the way of tumor metastasis inhibition [85, 86]. The vimentin expression in numerous types of cancer is shown by means of proteomic technologies that can support its central role the pathophysiology, diagnosis, prognosis, and therapy of cancer.

Integrins are a big family of heterodimeric transmembrane receptors and αvβ6 is an epithelial-specific one as a receptor for the extracellular matrix (ECM) proteins like fibronectin, vitronectin, tenascin, and the latency associated peptide (LAP) of TGF-β. Integrin αvβ6 is not expressed in healthy adult epithelia but is upregulated during wound healing and in cancer. The expression of the ß6 integrin is central for keeping the mesenchymal phenotype [87]. Integrin aVß6 has been verified to control invasive behavior of tumors, stop cell apoptosis, adjust the expression of matrix metalloproteases (MMPs), and activate TGF-ß1 [88]. In gastric cancer, non-small cell lung cancer (NSCLC), and liver tumors the expression of integrin aVß6 is completely linked to tumor clinical stage, differentiation, and tumor [89–91]. Another highlighted protein of cancer is glycogen synthasekinase-3 β (GSK-3 β) that is a proline-directed serine–threonine kinase. GSK3B is a multifunctional serine/threonine kinase which has been involved in several molecular pathways like embryonic development, cell differentiation, apoptosis, and insulin response [92]. Also, GSK3B is a crucial element of neuronal functions and is associated with main diseases regarding the CNS[12] [93]. Combined proteomics approaches indicated that GSK-3 β can trigger tumor genesis in several cancer types, such as mixed lineage leukemia, glioma, and oral cancer [94–96].

6 Cancer Stem Cells Metabolomics

Metabolome is a group of small metabolites (>1000 Dalton) such as metabolic intermediates, hormones, and some signaling molecules of a living organism [97, 98]. The major metabolite database (METLIN) is a source of metabolite information same as tandem mass spectrometry data.[13] In January 2007, a primary completed draft of the human metabolome was presented by University of Alberta and the University of Calgary named Human Metabolome Database (HMDB) [99]. A subgroup of metabolites which are different between cancerous cells and their counterpart noncancerous cells are called "oncometabolites" [100]. There are nine oncometabolites which are typically considered as oncometabolites in several tumors: 2-hydroxyglutarate (2-HG), glucose, fumarate, succinate, sarcosine, glutamine, asparagine, choline, and lactate [101]. CSCs are developing as important drivers of inter- and intra-tumoral heterogeneity and have their distinctive metabolic dependencies that are essential not only for exact bioenergetic/biosynthetic difficulties but also for supporting their functioning epigenetic traits, i.e., self-renewal, tumor initiation, and plasticity [102]. Metabolome as the final downstream product of all the omic layers can be the most representative of the biological phenotype specially in the way of understanding the complexity of tumor heterogeneity. Tumor heterogeneity as the mysterious characteristic of tumor is eventually in charge of tumor resistance to therapy and seed metastases. In fact, documentation of the driver nodes by main metabolite profiling as the differential one of CSC can shed a light on the self-renewal and tumor initiation and escape chemotherapy [103]. More than that, driver metabolites characterization for a large amount of modification within the CSC metabolomic sub-phenotypes might suggest new surprising chances for decreasing and developing tumor heterogeneity via metabolic targeting of CSC. Finding the main genetic and epigenetic alteration in consequence with main

[12] *CNS*: Central nervous system.

[13] *Tandem mass spectrometry*, also known as MS/MS or MS2, involves multiple steps of mass spectrometry selection, with some form of fragmentation occurring in between the stages.

oncometabolite alterations can represent a new valuable hallmark for cancer management. That means the metaboloepigenetic dimension of CSCs can improve the new metabostemness-targeting oncology drugs [104].

Urine covers metabolic marks of numerous biochemical pathways, so it can be a non-invasive ideal source for metabolomics analysis, especially involving diseases of the kidney and urinary system. It was shown that combination of three independent analytical techniques for detection of renal cell carcinoma (RCC) in urine of affected patients was efficient for lipophilic and hydrophilic urine metabolites detection [105]. Among 2000 mass spectral features which were sensed in the urine, some noteworthy components were suggested with discriminative value between RCC patients and controls. The metabolomics approaches disclose distinctive metabolic reprogramming in endothelial cells co-cultured with CSC and non-CSC prostate cancer cell subpopulations [106]. IBased on the non-targeted high-resolution liquid chromatography–mass spectrometry (LC-MS) 25 significantly altered metabolites were highlighted the most important ones including acetyl L-carnitine, NAD+, hypoxanthine, guanine, and oleamide.

7 Cancer Stem Cells Epigenomic

Epigenetics is mainly about the DNA methylation pattern as the heritable phenotype changes that do not contain alterations in the DNA sequence but is resulting in gene expression change (Fig. 3) [107]. The word epigenomics is equivalent to the study of the epigenome and mapping of the methylome which is changing the functioning

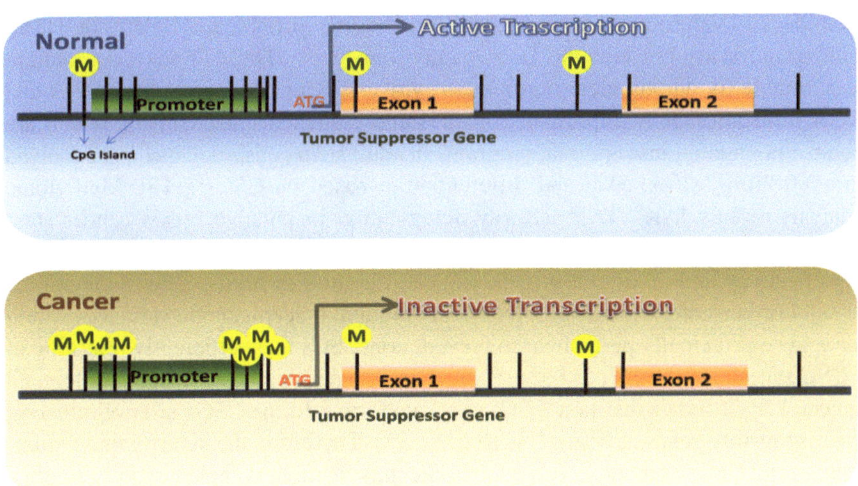

Fig. 3 The pattern of promoter methylation of tumor suppressor genes is different between normal and cancer cells

genome as a whole [108]. The methyl group attaches to the DNA mainly in CpG islands to activate or deactivate specific genes in the genome, so shifting the way that it behaves. There are two types of marks of an epigenomic compound, which are DNA methylation and histone modification. DNA methylation transfers the methyl groups (CH3) to the bases of DNA at specific receptor sites (CpG islands) to directly affect the DNA in a genome. The binding of the methyl group can control the gene on or off [109]. Another post-translational modifications histone modification process is histone modifications as the octamers that DNA wraps around them [110]. The epigenomic compounds can be passed on from one cell to another during the cell replication process to mark the next generation of cells. This means that they are heritable, but via cell meiosis and mitosis.

Genomes from several folks have already been sequenced and projects such as the HapMap highlighted the genetic variations in the whole human genome [111–113]. The ENCODE data describes function of human genome can indicated to functional elements in the genome and assisting the link of genes to complex disease phenotypes [114]. Such projects have the potential to help elucidate the information encoded by human genomes and aid in the treatment of diseases such as cancer. The environmental factors such as the exposure to chemical compounds during life, harmful radiation, smoking, and nutrition have definite impact on gene expression profile in human [115]. Thus, in the post-genomic era, studies of how human genes are regulated and the mechanisms that are implicated in this process are of major importance for our understanding of normal processes and diseased states. From the 1980s epigenetics have shown the potential of being talented targeted drugs that affect these mechanisms to treat diseases, especially cancers [116, 117]. Recently, the number of epidrugs affecting epigenetic mechanisms have increased and have been established to treat different types of cancer including Belinostat, Dacogen, Vidaza, Decitabine, DZNep,[14] Entinostat, Panobinostat, RG108, CP-4200, S110, Romidepsin, Valproic Acid, Vorinostat, Pyroxamide, Sirtinol, and Salermide [118]. This drug mostly prevents the DNA methyltransferase (DNMT) enzymes that are accountable for adding methyl groups to cytosines (CpG) placed in both DNA and RNA molecules. It is important to understand epigenetic mechanisms which are controlling main transcriptional programs in adult stem cells, such as those involved in controlling self-renewal and differentiation based on CSCs [119]. Methylome analysis reveals JAK-STAT pathway deregulation in putative breast cancer stem cells [120]. Silencing of SFRP1 by epigenetic mechanism can activate the canonical Wnt pathway and gives cell growth and proliferation in hepatocellular carcinoma [121]. Furthermore, while it was commonly thought that epigenetic variances would have to be principally permanent to prevent non-CSCs from efficiently returning to a CSC state, this concept has been recently challenged. A mouse model of colorectal cancer (CRC) shows that non-CSCs can transform into the CSCs after eliminating the responsible genes of being CSCs [122, 123]. Therefore, the determining parameters of the CSCs and non-CSCs epigenome may be more complex and plastic than previously believed. Recently it has been emphasized that epigenetic can be much

[14] *DZNep*: Deazaneplanocin A, histone methyltransferase inhibitor.

more important than genetic differences that are the underlying drivers for why CSCs are functionally different from their non-CSC counterparts. The most important epigenomic mechanisms are DNA methylation and/or demethylation, histone modification, non-coding RNA molecules, and nucleosome repositioning. DNA methylation can silence genes or non-coding SNA sequences through several mechanisms. Some literature has confirmed that methylation pattern alterations can happen in cancer development and progression [124]. The main change of methylation landscape occurs in embryonic stem cells (ESCs) during differentiation and cell fate decision [125]. Directional DNA methylation alteration and complex intermediate states complement lineage specificity in the adult hematopoietic compartment [126]. Actually, it is shown that in dissimilar tumor cells populations, distinct DNA methylation pattern can be the sign of crucial gene expressions like CD133/PROM1 expression in glioma stem cells and Sp1/myc and promoter methylation [127].

The breast cancer CSC genes CD44, CD133, and Musashi-1 (MSI1) are controlled by the methylation of promoter CpG regions, and hypomethylation can trigger these CSC genes in the more aggressive subtypes (triple-negative breast cancer) [128–130]. Comparing methylation profiles in invasive and non-invasive pancreatic cancer cells exposed an interesting connection between the methylation profile and the expression of central pathways, such as NF-kappa B signaling. Genes such as BIMP, TNFR, and CD49 were demethylated in the invasive form and methylated in the non-invasive form [131]. The driver genetic changes in CSCs are dependent on the class of enzymes known as DNA methyltransferases (DNMTs) that mediate the methylation reaction. The progress of leukemia was obstructed by DNMT1 functioning which induced re-expression of tumor suppressor genes, damaging CSC self-renewal, and reducing leukemia progression [132]. In DNMT1-knockout mice mammary stem and progenitor cell populations were abridged that DNMT1 have a critical role in the expansion and maintenance of mammary stem cells [133]. IL-6 enriched lung CSC via inhibition of cell cycle regulators via DNMT1 up regulation, which can be owed by methylation of TP53 and p21 (WAF1/CIP1) [134]. Pancreatic CSCs are in need of DNMT1 stemness properties as well and DNMT1 inhibition reprograms pancreatic CSCs by the miR-17-92 cluster [135].

The change of methylome[15] in CSCs and iPSCs[16] has been shown by high-resolution genome-wide mapping of methylation [136]. More than Dna methyl trasferase family, there are 10–11 translocation (TET) family proteins that dynamically contribute to DNA demethylation [137]. TET overexpression is important for tumor genesis process and its mutations can be detected in CRC, clear-cell renal cell carcinoma, and metastatic castration-resistant prostate cancer [138–140].

The microfluidic immune-precipitation followed by next generation sequencing technologies is enable scientists to check the epigenomics profiling of CSC and CTCs [141].

---

[15] *Methylome*: The set of nucleic acid methylation modifications in an organism's genome or in a particular cell.

[16] *iPSCs*: Induced pluripotent stem cells are a type of pluripotent stem cell that can be generated directly from adult cells.

8 Metaboloepigenomics of Cancer Stem Cells

In eukaryotic cells, some genetic change is can have link with other alterations during tumor genesis. It has been lately projected that metabolome can expressively affect the establishment and maintenance of epigenetic signatures of stemness and cancer [104]. So, DNA methylation and histone modification can be well observed as an integrated metaboloepigenetic dimension of CSCs, which we have recently termed cancer metabostemness. Chromatin modifying enzymes are dependent on cellular metabolites as sources of phosphate, acetyl, or methyl groups, and some of the energy providing metabolites like ATP, NAD^+, acetyl-CoA, and S-adenosylmethionine (SAM) so they are extremely needed by epigenetic enzymes that regulate DNA methylation and posttranslational modifications of histones [142–144]. That means, epigenetics works as a systematic connection between energy metabolism and gene expression since histones can act as bona fide metabolic sensors, altering the energy metabolites amount into stable patterns of gene expression. Not only whole but also localized variations in the altitudes of certain metabolites can provoke time-dependent variations in chromatin structure, consequently permitting the translation of a certain metabolic state into a specific histone map, pairing physiological states into metabolically driven changes of gene expression [145–148]. Oncometabolites change is a sign of cancer tissues, so it is rational to say that cancer-associated metabolic variations go along with chromatin-remodeling events. In spite of the fact that contributing association epigenetic modification has not been clearly established to tumors, indication is accruing on how some defects in genes encoding metabolic enzymes can imply on having impact on the DNMTs and histone modification enzymes. So, irregular accretion oncometabolites 2HG, succinate, fumarate, and sarcosine can trigger malignancy through DNA hypermethylation and histone demethylation [149, 150].

9 Future Perspectives

The knowledge of cancer biomarker discovery is moving toward more comprehensive and non-invasive tools. The omics can provide the comprehensive data over genetic mutation, protein expression, oncometabolites, and methylation pattern in tumors. Finding the linkage between this omics as the "trans-omic" analysis can reforming global biochemical networks across multiple omic layers. The multi-omic measurements and computational data integration make it possible. There is a big hope that in the near future technologies for connecting multi-omic data based on prior knowledge of biochemical interactions will highlight the most important biomarkers.

References

1. Moore KA, Lemischka IR. Stem cells and their niches. Science. 2006;311(5769):1880–5.
2. Reya T, Morrison SJ, Clarke MF, Weissman IL. Stem cells, cancer, and cancer stem cells. Nature. 2001;414(6859):105.
3. Ashkenazi R, Gentry SN, Jackson TL. Pathways to tumorigenesis—modeling mutation acquisition in stem cells and their progeny. Neoplasia. 2008;10(11):IN1–6.
4. Guo X, Wang X-F. Signaling cross-talk between TGF-β/BMP and other pathways. Cell Res. 2009;19(1):71.
5. Riquet M, Rivera C, Gibault L, Pricopi C, Mordant P, Badia A, et al. Lymphatic spread of lung cancer: anatomical lymph node chains unchained in zones. Rev Pneumol Clin. 2014; 70(1–2):16–25.
6. Gupta GP, Massagué J. Cancer metastasis: building a framework. Cell. 2006;127(4):679–95.
7. Ashworth T. A case of cancer in which cells similar to those in the tumours were seen in the blood after death. Med J Aust. 1869;14:146.
8. Racila E, Euhus D, Weiss AJ, Rao C, McConnell J, Terstappen LW, et al. Detection and characterization of carcinoma cells in the blood. Proc Natl Acad Sci. 1998;95(8):4589–94.
9. Khatami F, Larijani B, Tavangar SM. The presence of tumor extrachomosomal circular DNA (ecDNA) as a component of liquid biopsy in blood. Med Hypotheses. 2018;114:5–7.
10. Khatami F, Tavangar SM. Liquid biopsy in thyroid cancer: new insight. Int J Hematol Oncol Stem Cell Res. 2018;12(3):234–47.
11. Lapidot T, Sirard C, Vormoor J, Murdoch B, Hoang T, Caceres-Cortes J, et al. A cell initiating human acute myeloid leukaemia after transplantation into SCID mice. Nature. 1994; 367(6464):645.
12. Al-Hajj M, Wicha MS, Benito-Hernandez A, Morrison SJ, Clarke MF. Prospective identification of tumorigenic breast cancer cells. Proc Natl Acad Sci. 2003;100(7):3983–8.
13. Singh SK, Clarke ID, Terasaki M, Bonn VE, Hawkins C, Squire J, et al. Identification of a cancer stem cell in human brain tumors. Cancer Res. 2003;63(18):5821–8.
14. Ginestier C, Hur MH, Charafe-Jauffret E, Monville F, Dutcher J, Brown M, et al. ALDH1 is a marker of normal and malignant human mammary stem cells and a predictor of poor clinical outcome. Cell Stem Cell. 2007;1(5):555–67.
15. Kelly PN, Dakic A, Adams JM, Nutt SL, Strasser A. Tumor growth need not be driven by rare cancer stem cells. Science. 2007;317(5836):337.
16. Yoo M-H, Hatfield DL. The cancer stem cell theory: is it correct? Mol Cells. 2008;26(5):514.
17. Liu X, Fan D. The epithelial-mesenchymal transition and cancer stem cells: functional and mechanistic links. Curr Pharm Des. 2015;21(10):1279–91.
18. Ishiwata T. Cancer stem cells and epithelial-mesenchymal transition: novel therapeutic targets for cancer. Pathol Int. 2016;66(11):601–8.
19. Vailati-Riboni M, Palombo V, Loor JJ. What are omics sciences? In: Ametaj BN, editor. Periparturient diseases of dairy cows. Cham: Springer; 2017. p. 1–7.
20. Yugi K, Kubota H, Hatano A, Kuroda S. Trans-omics: how to reconstruct biochemical networks across multiple 'omic' layers. Trends Biotechnol. 2016;34(4):276–90.
21. Morozova O, Marra MA. Applications of next-generation sequencing technologies in functional genomics. Genomics. 2008;92(5):255–64.
22. Tomczak K, Czerwińska P, Wiznerowicz M. The cancer genome atlas (TCGA): an immeasurable source of knowledge. Contemp Oncol. 2015;19(1A):A68.
23. The future of cancer genomics. Nat Med. 2015;21:99.
24. Cerami E, Gao J, Dogrusoz U, Gross BE, Sumer SO, Aksoy BA, et al. The cBio cancer genomics portal: an open platform for exploring multidimensional cancer genomics data. Cancer Discov. 2012;2(5):401–4.
25. Regenbrecht CRA, Lehrach H, Adjaye J. Stemming cancer: functional genomics of cancer stem cells in solid tumors. Stem Cell Rev. 2008;4(4):319–28.

26. He X, Gonzalez V, Tsang A, Thompson J, Tsang TC, Harris DT. Differential gene expression profiling of CD34+ CD133+ umbilical cord blood hematopoietic stem progenitor cells. Stem Cells Dev. 2005;14(2):188–98.
27. Huang TS, Hsieh JY, Wu YH, Jen CH, Tsuang YH, Chiou SH, et al. Functional network reconstruction reveals somatic stemness genetic maps and dedifferentiation-like transcriptome reprogramming induced by GATA2. Stem Cells. 2008;26(5):1186–201.
28. Advani AS, Rodriguez C, Jin T, Jawde RA, Saber W, Baz R, et al. Increased C-kit intensity is a poor prognostic factor for progression-free and overall survival in patients with newly diagnosed AML. Leuk Res. 2008;32(6):913–8.
29. Boissel N, Leroy H, Brethon B, Philippe N, De Botton S, Auvrignon A, et al. Incidence and prognostic impact of c-Kit, FLT3, and Ras gene mutations in core binding factor acute myeloid leukemia (CBF-AML). Leukemia. 2006;20(6):965.
30. Stanulla M, Welte K, Hadam MR, Pietsch T. Coexpression of stem cell factor and its receptor c-Kit in human malignant glioma cell lines. Acta Neuropathol. 1995;89(2):158–65.
31. Musci TJ, Amaya E, Kirschner MW. Regulation of the fibroblast growth factor receptor in early Xenopus embryos. Proc Natl Acad Sci. 1990;87(21):8365–9.
32. Nishita Y, Sado T, Yoshida I, Takagi N. Effect of CpG methylation on expression of the mouse imprinted gene Mest. Gene. 1999;226(2):199–209.
33. Dennis G, Sherman BT, Hosack DA, Yang J, Gao W, Lane HC, et al. DAVID: database for annotation, visualization, and integrated discovery. Genome Biol. 2003;4(9):R60.
34. Kanehisa M, Goto S. KEGG: Kyoto encyclopedia of genes and genomes. Nucleic Acids Res. 2000;28(1):27–30.
35. Karamboulas C, Ailles L. Developmental signaling pathways in cancer stem cells of solid tumors. Biochim Biophys Acta. 2013;1830(2):2481–95.
36. Sell S. Cancer stem cells and differentiation therapy. Tumor Biol. 2006;27(2):59–70.
37. Zhang T, Xu J, Deng S, Zhou F, Li J, Zhang L, et al. Core signaling pathways in ovarian cancer stem cell revealed by integrative analysis of multi-marker genomics data. PLoS One. 2018;13(5):e0196351.
38. Matsumura N, Yamanoi K, Murphy SK, Hamanishi J, Abiko K, Yamaguchi K, et al. Abstract A38: identification through functional genomics screening of factors whose downregulation enhances the side population in ovarian cancer. Clin Cancer Res. 2016;22(2 Suppl):A38.
39. Chen Y-A, Lu C-Y, Cheng W-F, Ho H-N, Cheng H-F. Propagating the ovarian cancer stem cells by reprogramming the tumor cells. FASEB J. 2017;31(1_suppl):395.1.
40. Cortina C, Turon G, Stork D, Hernando-Momblona X, Sevillano M, Aguilera M, et al. A genome editing approach to study cancer stem cells in human tumors. EMBO Mol Med. 2017;9(7):869–79.
41. Lohr JG, Adalsteinsson VA, Cibulskis K, Choudhury AD, Rosenberg M, Cruz-Gordillo P, et al. Whole-exome sequencing of circulating tumor cells provides a window into metastatic prostate cancer. Nat Biotechnol. 2014;32(5):479.
42. Premasekharan G, Gilbert E, Okimoto RA, Hamirani A, Lindquist KJ, Ngo VT, et al. An improved CTC isolation scheme for pairing with downstream genomics: demonstrating clinical utility in metastatic prostate, lung and pancreatic cancer. Cancer Lett. 2016;380(1):144–52.
43. Gupta S, Li J, Kemeny G, Bitting RL, Beaver J, Somarelli JA, et al. Whole genomic copy number alterations in circulating tumor cells from men with abiraterone or enzalutamide-resistant metastatic castration-resistant prostate cancer. Clin Cancer Res. 2017;23(5):1346–57.
44. Singhal U, Wang Y, Henderson J, Niknafs YS, Qiao Y, Gursky A, et al. Multigene profiling of CTCs in mCRPC identifies a clinically relevant prognostic signature. Mol Cancer Res. 2018;16(4):643–54.
45. Castro-Giner F, Scheidmann MC, Aceto N. Beyond enumeration: functional and computational analysis of circulating tumor cells to investigate cancer metastasis. Front Med. 2018;5:34.
46. Carter L, Rothwell DG, Mesquita B, Smowton C, Leong HS, Fernandez-Gutierrez F, et al. Molecular analysis of circulating tumor cells identifies distinct copy-number pro-

files in patients with chemosensitive and chemorefractory small-cell lung cancer. Nat Med. 2017;23(1):114.
47. Hou Y, Song L, Zhu P, Zhang B, Tao Y, Xu X, et al. Single-cell exome sequencing and monoclonal evolution of a JAK2-negative myeloproliferative neoplasm. Cell. 2012;148(5):873–85.
48. Lodato MA, Woodworth MB, Lee S, Evrony GD, Mehta BK, Karger A, et al. Somatic mutation in single human neurons tracks developmental and transcriptional history. Science. 2015;350(6256):94–8.
49. Zafar H, Wang Y, Nakhleh L, Navin N, Chen K. Monovar: single-nucleotide variant detection in single cells. Nat Methods. 2016;13(6):505.
50. Gawad C, Koh W, Quake SR. Single-cell genome sequencing: current state of the science. Nat Rev Genet. 2016;17(3):175.
51. Wang Y, Navin NE. Advances and applications of single-cell sequencing technologies. Mol Cell. 2015;58(4):598–609.
52. Garvin T, Aboukhalil R, Kendall J, Baslan T, Atwal GS, Hicks J, et al. Interactive analysis and assessment of single-cell copy-number variations. Nat Methods. 2015;12(11):1058.
53. Hou Y, Fan W, Yan L, Li R, Lian Y, Huang J, et al. Genome analyses of single human oocytes. Cell. 2013;155(7):1492–506.
54. Knouse KA, Wu J, Amon A. Assessment of megabase-scale somatic copy number variation using single-cell sequencing. Genome Res. 2016;26(3):376–84.
55. McConnell MJ, Lindberg MR, Brennand KJ, Piper JC, Voet T, Cowing-Zitron C, et al. Mosaic copy number variation in human neurons. Science. 2013;342(6158):632–7.
56. Anderson NL, Anderson NG. Proteome and proteomics: new technologies, new concepts, and new words. Electrophoresis. 1998;19(11):1853–61.
57. Blackstock WP, Weir MP. Proteomics: quantitative and physical mapping of cellular proteins. Trends Biotechnol. 1999;17(3):121–7.
58. James P. Protein identification in the post-genome era: the rapid rise of proteomics. Q Rev Biophys. 1997;30(4):279–331.
59. Swinbanks D. Australia backs innovation, shuns telescope. Nature. 1995;378(6558):653.
60. Bi B, Li F, Guo J, Li C, Jing R, Lv X, et al. Label-free quantitative proteomics unravels the importance of RNA processing in glioma malignancy. Neuroscience. 2017;351:84–95.
61. Zhang W, Liu T, Dong H, Bai H, Tian F, Shi Z, et al. Synthesis of a highly azide-reactive and thermosensitive biofunctional reagent for efficient enrichment and large-scale identification of O-GlcNAc proteins by mass spectrometry. Anal Chem. 2017;89(11):5810–7.
62. Webber JP, Stone TC, Katilius E, Smith BC, Gordon B, Mason MD, et al. Proteomics analysis of cancer exosomes using a novel modified aptamer-based array (SOMAscan™) platform. Mol Cell Proteomics. 2014;13(4):1050–64.
63. Zaccaria A, Bouamrani A, Selek L, El Atifi M, Hesse AM. Juhem Al, et al. a micro-silicon chip for in vivo cerebral imprint in monkey. ACS Chem Neurosci. 2013;4(3):385–92.
64. Ji L, Jayachandran G, Roth JA. High throughput profiling of serum phosphoproteins/peptides using the SELDI-TOF-MS platform. Methods Mol Biol. 2012;818:199–216.
65. Li S, Sack R, Vijmasi T, Sathe S, Beaton A, Quigley D, et al. Antibody protein array analysis of the tear film cytokines. Optom Vis Sci. 2008;85(8):653.
66. Gupta S, Manubhai K, Kulkarni V, Srivastava S. An overview of innovations and industrial solutions in protein microarray technology. Proteomics. 2016;16(8):1297–308.
67. Cox J, Hein MY, Luber CA, Paron I, Nagaraj N, Mann M. Accurate proteome-wide label-free quantification by delayed normalization and maximal peptide ratio extraction, termed MaxLFQ. Mol Cell Proteomics. 2014;13(9):2513–26.
68. Kanojia D, Zhou W, Zhang J, Jie C, Lo PK, Wang Q, et al. Proteomic profiling of cancer stem cells derived from primary tumors of HER2/Neu transgenic mice. Proteomics. 2012; 12(22):3407–15.
69. Nie S, McDermott SP, Deol Y, Tan Z, Wicha MS, Lubman DM. A quantitative proteomics analysis of MCF7 breast cancer stem and progenitor cell populations. Proteomics. 2015;15(22):3772–83.

70. Lee E-K, Cho H, Kim C-W. Proteomic analysis of cancer stem cells in human prostate cancer cells. Biochem Biophys Res Commun. 2011;412(2):279–85.
71. Zhong Z, Mao S, Lin H, Lin J-M, Lin J. Comparative proteomics of cancer stem cells in osteosarcoma using ultra-high-performance liquid chromatography and Orbitrap fusion mass spectrometer. Talanta. 2018;178:362–8.
72. Ko CH, Cheng CF, Lai CP, Tzu TH, Chiu CW, Lin MW, et al. Differential proteomic analysis of cancer stem cell properties in hepatocellular carcinomas by isobaric tag labeling and mass spectrometry. J Proteome Res. 2013;12(8):3573–85.
73. Morisaki T, Yashiro M, Kakehashi A, Inagaki A, Kinoshita H, Fukuoka T, et al. Comparative proteomics analysis of gastric cancer stem cells. PLoS One. 2014;9(11):e110736.
74. van Houdt WJ, Emmink BL, Pham TV, Piersma SR, Verheem A, Vries R, et al. Comparative proteomics of colon cancer stem cells and differentiated tumor cells identifies BIRC6 as a potential therapeutic target. Mol Cell Proteomics. 2011;10(12):M111.011353.
75. Scatena R, Bottoni P, Giardina B. Circulating tumour cells and cancer stem cells: a role for proteomics in defining the interrelationships between function, phenotype and differentiation with potential clinical applications. Biochimica et Biophysica Acta (BBA)-reviews on. Cancer. 2013;1835(2):129–43.
76. Halvey PJ, Zhang B, Coffey RJ, Liebler DC, Slebos RJ. Proteomic consequences of a single gene mutation in a colorectal cancer model. J Proteome Res. 2011;11(2):1184–95.
77. Fang D, Kim Y, Lee C, Aggarwal S, McKinnon K, Mesmer D, et al. Expansion of CD133+ colon cancer cultures retaining stem cell properties to enable cancer stem cell target discovery. Br J Cancer. 2010;102(8):1265.
78. Yen L, Cao Z, Wu X, Ingalla ER, Baron C, Young LJ, et al. Loss of Nrdp1 enhances ErbB2/ErbB3–dependent breast tumor cell growth. Cancer Res. 2006;66(23):11279–86.
79. Hermiston ML, Wong MH, Gordon JI. Forced expression of E-cadherin in the mouse intestinal epithelium slows cell migration and provides evidence for nonautonomous regulation of cell fate in a self-renewing system. Genes Dev. 1996;10(8):985–96.
80. Chu Y-S, Eder O, Thomas WA, Simcha I, Pincet F, Ben-Ze'ev A, et al. Prototypical type I E-cadherin and type II cadherin-7 mediate very distinct adhesiveness through their extracellular domains. J Biol Chem. 2006;281(5):2901–10.
81. Berx G, Van Roy F. Involvement of members of the cadherin superfamily in cancer. Cold Spring Harb Perspect Biol. 2009;1(6):a003129.
82. Mor-Vaknin N, Punturieri A, Sitwala K, Markovitz DM. Vimentin is secreted by activated macrophages. Nat Cell Biol. 2003;5(1):59.
83. Garg A, Barnes PF, Porgador A, Roy S, Wu S, Nanda JS, et al. Vimentin expressed on Mycobacterium tuberculosis-infected human monocytes is involved in binding to the NKp46 receptor. J Immunol. 2006;177(9):6192–8.
84. Chong WP, Zhou J, Law HK, Tu W, Lau YL. Natural killer cells become tolerogenic after interaction with apoptotic cells. Eur J Immunol. 2010;40(6):1718–27.
85. Halfteck GG, Elboim M, Gur C, Achdout H, Ghadially H, Mandelboim O. Enhanced in vivo growth of lymphoma tumors in the absence of the NK-activating receptor NKp46/NCR1. J Immunol. 2009;182(4):2221–30.
86. Lakshmikanth T, Burke S, Ali TH, Kimpfler S, Ursini F, Ruggeri L, et al. NCRs and DNAM-1 mediate NK cell recognition and lysis of human and mouse melanoma cell lines in vitro and in vivo. J Clin Invest. 2009;119(5):1251–63.
87. Annes JP, Chen Y, Munger JS, Rifkin DB. Integrin αVβ6-mediated activation of latent TGF-β requires the latent TGF-β binding protein-1. J Cell Biol. 2004;165(5):723–34.
88. Manti S, Marseglia L, D'Angelo G, Filippelli M, Cuppari C, Gitto E, et al. Portal hypertension as immune mediate disease. Hepat Mon. 2014;14(6):e18625.
89. Bandyopadhyay A, Raghavan S. Defining the role of integrin αvβ6 in cancer. Curr Drug Targets. 2009;10(7):645–52.

90. Kong L, Sun J, Zhang C. The expression of integrinαvβ6 and MMP-9 in gastric cancer and the correlation with clinicopathologic characteristic. Xi Bao Yu Fen Zi Mian Yi Xue Za Zhi. 2011;27(1):92–4.
91. Patsenker E, Wilkens L, Banz V, Österreicher C, Weimann R, Eisele S, et al. The αvβ6 integrin is a highly specific immunohistochemical marker for cholangiocarcinoma. J Hepatol. 2010;52(3):362–9.
92. Jacobs KM, Bhave SR, Ferraro DJ, Jaboin JJ, Hallahan DE, Thotala D. GSK-3β: a Bifunctional role in cell death pathways. Int J Cell Biol. 2012;2012:930710.
93. Luo J. The role of GSK3beta in the development of the central nervous system. Front Biol. 2012;7(3):212–20.
94. Lee CF, Griffiths S, Rodríguez-Suárez E, Pierce A, Unwin RD, Jaworska E, et al. Assessment of downstream effectors of BCR/ABL protein tyrosine kinase using combined proteomic approaches. Proteomics. 2010;10(18):3321–42.
95. da Gama Fischer JDS, Liao L, Carvalho PC, Barbosa VC, Domont GB, da Costa Carvalho MDG, et al. Dynamic proteomic overview of glioblastoma cells (A172) exposed to perillyl alcohol. J Proteome. 2010;73(5):1018–27.
96. Mishra R. Glycogen synthase kinase 3 beta: can it be a target for oral cancer. Mol Cancer. 2010;9(1):144.
97. Oliver SG, Winson MK, Kell DB, Baganz F. Systematic functional analysis of the yeast genome. Trends Biotechnol. 1998;16(9):373–8.
98. Griffin JL, Vidal-Puig A. Current challenges in metabolomics for diabetes research: a vital functional genomic tool or just a ploy for gaining funding? Physiol genomics. 2017;34:1–5.
99. Wishart DS, Jewison T, Guo AC, Wilson M, Knox C, Liu Y, et al. HMDB 3.0—the human metabolome database in 2013. Nucleic Acids Res. 2012;41(D1):D801–D7.
100. Sciacovelli M, Frezza C. Oncometabolites: unconventional triggers of oncogenic signalling cascades. Free Radic Biol Med. 2016;100:175–81.
101. Wishart DS, Mandal R, Stanislaus A, Ramirez-Gaona M. Cancer metabolomics and the human metabolome database. Meta. 2016;6(1):10.
102. Cuyàs E, Verdura S, Fernández-Arroyo S, Bosch-Barrera J, Martin-Castillo B, Joven J, et al. Metabolomic mapping of cancer stem cells for reducing and exploiting tumor heterogeneity. Oncotarget. 2017;8(59):99223–36.
103. Tabassum DP, Polyak K. Tumorigenesis: it takes a village. Nat Rev Cancer. 2015;15(8):473.
104. Menendez JA. The metaboloepigenetic dimension of cancer stem cells: evaluating the market potential for new metabostemness-targeting oncology drugs. Curr Pharm Des. 2015; 21(25):3644–53.
105. Kind T, Tolstikov V, Fiehn O, Weiss RH. A comprehensive urinary metabolomic approach for identifying kidney cancer. Anal Biochem. 2007;363(2):185–95.
106. Jayaraman A, Kumar P, Marin S, de Atauri P, Mateo F, Thomson TM, et al. Untargeted metabolomics reveals distinct metabolic reprogramming in endothelial cells co-cultured with CSC and non-CSC prostate cancer cell subpopulations. PLoS One. 2018;13(2):e0192175.
107. Bird A. DNA methylation patterns and epigenetic memory. Genes Dev. 2002;16(1):6–21.
108. Epigenomics EV. Mapping the methylome. Cell Cycle. 2006;5(2):155–8.
109. Jones PA, Takai D. The role of DNA methylation in mammalian epigenetics. Science. 2001; 293(5532):1068–70.
110. Seligson DB, Horvath S, Shi T, Yu H, Tze S, Grunstein M, et al. Global histone modification patterns predict risk of prostate cancer recurrence. Nature. 2005;435(7046):1262.
111. Levy S, Sutton G, Ng PC, Feuk L, Halpern AL, Walenz BP, et al. The diploid genome sequence of an individual human. PLoS Biol. 2007;5(10):e254.
112. Wang J, Wang W, Li R, Li Y, Tian G, Goodman L, et al. The diploid genome sequence of an Asian individual. Nature. 2008;456(7218):60.
113. Consortium IH. A haplotype map of the human genome. Nature. 2005;437(7063):1299.
114. Consortium EP. Identification and analysis of functional elements in 1% of the human genome by the ENCODE pilot project. Nature. 2007;447(7146):799.

115. López-Maury L, Marguerat S, Bähler J. Tuning gene expression to changing environments: from rapid responses to evolutionary adaptation. Nat Rev Genet. 2008;9(8):583.
116. Wilson VL, Jones PA, Momparler RL. Inhibition of DNA methylation in L1210 leukemic cells by 5-aza-2′-deoxycytidine as a possible mechanism of chemotherapeutic action. Cancer Res. 1983;43(8):3493–6.
117. Jones PA. Altering gene expression with 5-azacytidine. Cell. 1985;40(3):485.
118. Jones PA, Issa J-PJ, Baylin S. Targeting the cancer epigenome for therapy. Nat Rev Genet. 2016;17(10):630.
119. Balic M, Schwarzenbacher D, Stanzer S, Heitzer E, Auer M, Geigl JB, et al. Genetic and epigenetic analysis of putative breast cancer stem cell models. BMC Cancer. 2013;13(1):358.
120. Hernandez-Vargas H, Ouzounova M, Le Calvez-Kelm F, Lambert M-P, McKay-Chopin S, Tavtigian SV, et al. Methylome analysis reveals Jak-STAT pathway deregulation in putative breast cancer stem cells. Epigenetics. 2011;6(4):428–39.
121. Kaur P, Mani S, Cros M-P, Scoazec J-Y, Chemin I, Hainaut P, et al. Epigenetic silencing of sFRP1 activates the canonical Wnt pathway and contributes to increased cell growth and proliferation in hepatocellular carcinoma. Tumor Biol. 2012;33(2):325–36.
122. de Sousa e Melo F, Kurtova AV, Harnoss JM, Kljavin N, Hoeck JD, Hung J, et al. A distinct role for Lgr5+ stem cells in primary and metastatic colon cancer. Nature. 2017;543(7647):676.
123. Shimokawa M, Ohta Y, Nishikori S, Matano M, Takano A, Fujii M, et al. Visualization and targeting of LGR5+ human colon cancer stem cells. Nature. 2017;545(7653):187.
124. Feinberg AP, Vogelstein B. Hypomethylation distinguishes genes of some human cancers from their normal counterparts. Nature. 1983;301(5895):89.
125. Gifford CA, Ziller MJ, Gu H, Trapnell C, Donaghey J, Tsankov A, et al. Transcriptional and epigenetic dynamics during specification of human embryonic stem cells. Cell. 2013; 153(5):1149–63.
126. Hodges E, Molaro A, Dos Santos CO, Thekkat P, Song Q, Uren PJ, et al. Directional DNA methylation changes and complex intermediate states accompany lineage specificity in the adult hematopoietic compartment. Mol Cell. 2011;44(1):17–28.
127. Gopisetty G, Xu J, Sampath D, Colman H, Puduvalli V. Epigenetic regulation of CD133/PROM1 expression in glioma stem cells by Sp1/myc and promoter methylation. Oncogene. 2013;32(26):3119.
128. Baba T, Convery P, Matsumura N, Whitaker R, Kondoh E, Perry T, et al. Epigenetic regulation of CD133 and tumorigenicity of CD133+ ovarian cancer cells. Oncogene. 2009;28(2):209.
129. Kagara N, Huynh KT, Kuo C, Okano H, Sim MS, Elashoff D, et al. Epigenetic regulation of cancer stem cell genes in triple-negative breast cancer. Am J Pathol. 2012;181(1):257–67.
130. Yi JM, Tsai H-C, Glöckner SC, Lin S, Ohm JE, Easwaran H, et al. Abnormal DNA methylation of CD133 in colorectal and glioblastoma tumors. Cancer Res. 2008;68(19):8094–103.
131. Sun L, Mathews LA, Cabarcas SM, Zhang X, Yang A, Zhang Y, et al. Epigenetic regulation of SOX9 by the NF-κB signaling pathway in pancreatic cancer stem cells. Stem Cells. 2013;31(8):1454–66.
132. Trowbridge JJ, Sinha AU, Zhu N, Li M, Armstrong SA, Orkin SH. Haploinsufficiency of Dnmt1 impairs leukemia stem cell function through derepression of bivalent chromatin domains. Genes Dev. 2012;26(4):344–9.
133. Pathania R, Ramachandran S, Elangovan S, Padia R, Yang P, Cinghu S, et al. DNMT1 is essential for mammary and cancer stem cell maintenance and tumorigenesis. Nat Commun. 2015;6:6910.
134. Liu CC, Lin JH, Hsu TW, Su K, Li AFY, Hsu HS, et al. IL-6 enriched lung cancer stem-like cell population by inhibition of cell cycle regulators via DNMT1 upregulation. Int J Cancer. 2015;136(3):547–59.
135. Zagorac S, Alcala S, Bayon GF, Kheir TB, Schoenhals M, González-Neira A, et al. DNMT1 inhibition reprograms pancreatic cancer stem cells via upregulation of the miR-17-92 cluster. Cancer Res. 2016;76(15):4546–58.

136. Baylin SB, Jones PA. A decade of exploring the cancer epigenome—biological and translational implications. Nat Rev Cancer. 2011;11(10):726.
137. Ito S, D'alessio AC, Taranova OV, Hong K, Sowers LC, Zhang Y. Role of Tet proteins in 5mC to 5hmC conversion, ES-cell self-renewal and inner cell mass specification. Nature. 2010;466(7310):1129.
138. Gu T-P, Guo F, Yang H, Wu H-P, Xu G-F, Liu W, et al. The role of Tet3 DNA dioxygenase in epigenetic reprogramming by oocytes. Nature. 2011;477(7366):606.
139. Sato Y, Yoshizato T, Shiraishi Y, Maekawa S, Okuno Y, Kamura T, et al. Integrated molecular analysis of clear-cell renal cell carcinoma. Nat Genet. 2013;45(8):860.
140. Nickerson ML, Im KM, Misner KJ, Tan W, Lou H, Gold B, et al. Somatic alterations contributing to metastasis of a castration-resistant prostate cancer. Hum Mutat. 2013;34(9):1231–41.
141. Zhu Y. Microfluidic technology for low-input epigenomic analysis. Blacksburg: Virginia Tech; 2018.
142. Donohoe DR, Bultman SJ. Metaboloepigenetics: interrelationships between energy metabolism and epigenetic control of gene expression. J Cell Physiol. 2012;227(9):3169–77.
143. Lu C, Thompson CB. Metabolic regulation of epigenetics. Cell Metab. 2012;16(1):9–17.
144. Katada S, Imhof A, Sassone-Corsi P. Connecting threads: epigenetics and metabolism. Cell. 2012;148(1–2):24–8.
145. Martinez-Pastor B, Cosentino C, Mostoslavsky R. A tale of metabolites: the cross-talk between chromatin and energy metabolism. Cancer Discov. 2013;3(5):497–501.
146. Salehzadeh-Yazdi A, Asgari Y, Saboury AA, Masoudi-Nejad A. Computational analysis of reciprocal association of metabolism and epigenetics in the budding yeast: a genome-scale metabolic model (GSMM) approach. PLoS One. 2014;9(11):e111686.
147. Carey BW, Finley LW, Cross JR, Allis CD, Thompson CB. Intracellular α-ketoglutarate maintains the pluripotency of embryonic stem cells. Nature. 2015;518(7539):413.
148. Teslaa T, Teitell MA. Pluripotent stem cell energy metabolism: an update. EMBO J. 2015; 34(2):138–53.
149. Goding CR, Pei D, Lu X. Cancer: pathological nuclear reprogramming? Nat Rev Cancer. 2014;14(8):568.
150. Raabe EH, Eberhart CG. Methylome alterations "mark" new therapeutic opportunities in glioblastoma. Cancer Cell. 2012;22(4):417–8.

From OMICs to Ethics: Points to Start the Debate

Leila Afshar

Abstract As any other field of the medicine which is experiencing fast and broad development, the OMICs methods face new challenges in terms of ethical and policy issues. Therefore in this chapter, by reviewing the greatest impact of these innovations on medicine and its domains, the main ethical issues in therapeutic relationship and clinical researches will be interpreted in terms of OMICs ethics. Therefore after some introducing paragraphs, this chapter consists of three main parts and their subcategories: the ethical foundations of therapeutic relationship and their meaning in OMICs era, the principles of clinical research ethics and their interpretation in OMICs field, and finally the social and policy points in OMICs practice.

Keywords OMICs · Biomedical ethics · Research ethics

1 Introduction

The term OMICs refers to technologies and methods used to explore the roles, relationships, and actions of the various types of a large family of cellular molecules, such as genes, proteins, or small metabolites, which together make up an organism. Few scientific fields and technologies have experienced such a development and significant increase in utility as OMICs science. However, there is a question on the ethical aspects of these developments. Are these developments and their applications raise new ethical issues? This means that we should consider whether our current ethical framework in the field of biomedicine provides genuine answers to our questions or we need novel approaches to keep pace with these rapid developments.

In this chapter, I try to introduce current approaches in biomedical ethics and will argue that based on their inefficiency in OMICs field, we need a different approach to develop a sound OMICs ethics; in other words, it is necessary to reset the parameters that frame this area of biomedical ethics. This means that our major

L. Afshar (✉)
Medical Ethics Department, Shahid Beheshti University of Medical Sciences, Tehran, Iran
e-mail: leilaafshar@sbmu.ac.ir

B. Arjmand (ed.), *Genomics, Proteomics, and Metabolomics*, Stem Cell Biology and Regenerative Medicine, https://doi.org/10.1007/978-3-030-27727-7_10

focuses of research ethics in this area, although are necessary and useful, are not sufficient and need some more in-depth considerations. First, what can be considered as a necessary condition for any theory of OMICs ethics will be outlined. Second, by reviewing the main ethical theories in the field of biomedicine, it would be suggested that the principle-based approach, which is currently dominant in much field of biomedical ethics and is very useful in the field of research, is not sufficient for analyzing the complex issues in the field of OMICs because of the primacy it accords to individuality and the idea of non-interference. And finally some areas in this fields that seems need more attention will briefly outline from an ethical perspective.

2 Where OMICs Meets the Ethics: Conceptual Foundations

It is proved that new features of the OMICs practice are striking and will have effects on social normative institution such as ethics and governmental policies. At least the scale and pace of the OMICs practice, its technology and methods, will lead a paradigm shift in theory development which allows to practice hypothesis-free and hypothesis-generating researches in this regard [1].

Therefore, it seems that there is a pressing need to address arising questions, but first we must look at the way biomedical ethics traversed and main ethical theories that support it, focusing on three main category: first, the shift of decision-making process in therapeutic relationship and its effect on physician–patient relationship; second, research ethics; finally and the most important one, the interaction between medicine as a profession and the society, keeping in mind the social and policy issues. However, it is necessary to mention here that the focus is on the contemporary biomedical ethics mainstream, and this does not mean that other important traditions of the medical ethics were neglected. This selection is because of the dominancy of this approach in framing the regulation and guidelines.

2.1 The Ethics of Therapeutic Relationships

The term “medical ethics” was not coined until 1803, when Thomas Percival, an English physician, introduced it in his book [2, 3]. He defined medical ethics as a description of physician’s duties toward their patients and society. However, the concept of medical ethics has developed over millennia in many different cultural, medical, and religious contexts. These traditions presented codes and guidelines for professionals, which may contain rules about how to behave—medical etiquette—and how to make ethical decisions as a good doctor in therapeutic relationships—medical ethics.

Beyond formal codes which try to establish ethical standards, themes from religious texts and the work of moral philosophers have shaped thinking about the profession of medicine and the moral duties of physicians.

Over the centuries, these codices demonstrated the fixed and well-known duties of the physician to the patients, in the condensed form of the physician's pledge, or oath, from Maimonides or Hippocrates, which are being continued through the modern ones.

Of particular relevance to our topic of OMICs ethics, some of these universal traditional promises such as confidentiality play a great role in the patient–physician relationship and also in data protection in researches. However, less attention has been paid to some other aspects of this relationship, such as acknowledging the patients' value and empowering them [4].

Keep in mind that the current framework of biomedical ethics mainly aroused as a reaction to crimes against humanity and scandals of medical malpractice. Therefore, the key concept of it is protectionism, and the object of this protection paradigm is human dignity. Values such as individual free will and the right to self-determination are the core notion of this paradigm. Therefore, autonomy became the central concept and the requirements of respecting autonomy have become the leading principle in Western world.

However, in principle-based approach to biomedical ethics, classical principles consist of the four well-known clusters of norms that were first presented by Beauchamp and Childress in their textbook, in 1979 [5–8].

1. Respect for autonomy can be stated that: rational individuals should be permitted to self-determination. This means respecting the decision-making capacities of autonomous persons, and establishing the requirements of voluntariness and of consent.
2. Non-maleficence or avoiding the causation of harm is a representation of the most famous and perhaps the most quoted maxims of medicine: *primum non nocere*
3. Beneficence provides norms for benefiting the patients and balancing benefits against risks and costs.
4. Justice is a norm for distributing benefits, risks, and costs fairly.

As making moral decisions is always a difficult and stressful task, these principles will guide the relationship between the individual patient and the physician in clinical practice, as well as the relationships in clinical research. However in recent decades, some ethical theorists have turned away to other approaches. Some emphasized on the importance of character as the source of moral action, whereas others pointed the central role of shared concerns and the importance of social relationships.

Based on the above-mentioned theoretical and historical foundations, and considering the influences of OMICs practice in the field of medicine, the rules that governing the therapeutic relations should be verified.

It's noteworthy to mention that the most significant impact of OMICs practice in medicine is a new paradigm of personalized medicine, which leads to a trend of projects promising high utility tests for the prediction of disease risks and early interventions. Along with this trend, there are growing number of direct-to-consumer tests and interventions which will directly affect the patient's decisions [9].

Considering these issues shows us that we need a new paradigm in physician–patient relationship that necessarily did not match the dominancy of respecting autonomy [10]. A health care provider should choose the most significant cluster of biomedical principles in this regard, which seems that are non-maleficence and beneficence in a more comprehensive sense. And also the principle of distributive justice in a societal level should be considered. The challenges and limitations of OMICs need to be carefully addressed, especially when it comes to predictive ability of them which could be tricky, for example, the relationship between association and causation, and the false-positive results because over-testing could result in overdiagnosis and overtreatment [11], and it should be noted that it is the responsibility of the physician to educate the patients and empower them to have informed decision.

2.2 Research Ethics in the OMICs Era

The history of medical research witnessed tremendous controversy about the ethics of clinical research. During the time, different topics were the issue of debate and alteration such as: using placebo, research on children and mentally incapable patients as vulnarable populations, the researcher's obligation to participants after finishing trials and the financial conflict of interest in externally sponsored researches. Currently the ethical challenges of international collaboration in research gained much attention [12]. These controversies have forced the reexamination of fundamental issues long deemed settled and produced much of the ethical guidance for clinical research.

Most of guidelines were a response to a specific subject, and therefore tend to focus on a scandal. The Nuremberg Code directly addressed the atrocities of the Nazi physicians [13]; the Belmont Report was a response to the Tuskegee Syphilis Study and other scandals (http://ohsr.od.nih.gov/guidelines/belmont.html); and so on. Therefore, regulatory guidance tends not to mention the overall ethics of research but to have a specific practical purpose. For example, conflict of interest was addressed in 2004 version of the Declaration of Helsinki by proposing the disclosure strategy, which requires that the potential research participants should be adequately informed about "any possible conflict of interest" and that also should be declared in the publications [14], or the 2013 version of it focuses on the use of placebo and its conditions [15].

Because of the deficiencies of existing research ethics guidance, a broader, systematic, and comprehensive framework was proposed; the framework incorporated concerns that overlap in the existing guidance and organized them into a coherent whole [16]. This framework configured eight principles that identify necessary considerations for justifying research as an ethical activity: collaborative partnership, social value, scientific validity, fair participant selection, favorable risk–benefit ratio, independent review, informed consent, and respect for participants.

Translating these principles to OMICs, researches will show us a need for focusing on special and more important issues than before.

Collaborative partnership: The first principle reveals us the importance of collaborative partnership in research. The principle recognizes that if the clinical research is serving a social good, meaning enhancing the health of the people, these people should collaborate in the research endeavor to ensure that the results of the research have long-lasting effects and will influence the policy makers [17, 18]. In new research domains such as OMICs, this needs formal and informal mechanisms and several benchmarks. The most important one could be a shared responsibility to find the health problem which needs the OMICs interventions based on the effect of the problem and the community resources. This will develop the necessary infrastructure of the research.

Social value of the research reminds us that the OMICs research itself is not the end. They have instrumental values to generate knowledge and improve health [19]. Without considering the social value, OMICs research will have no good reason and would be a vast of resources. Emphasis on protectionism and scientific development should not undermine the importance of the sound justification for doing a research. OMICs researches that address a problem with little relevance to a real health problem lack the social value critria, as if the finding that never could be practically implemented.

Scientific validity stresses that producing valid and reliable data is an ethical obligation, and this means that the research design and methods, scientifically and statistically, should be sound and the research must have clear objectives, unbiased and reliable outcome measures. In the field of OMICs that we would have innovative designs and protocols, any deviations from the standard protocols must plausibly be justified. Another important issue in this regard is practical feasibility of the research findings which again reminds us that OMICs researchers should think beyond their lab.

Fair subject selection and *favorable risk–benefit* ratio are two other important principles in clinical researches. In the OMICs research, these principles should be considered more in-depth and comprehensive, especially, in first, in human and phase 1 safety studies and even when the risks of a new intervention are blur or outweigh the potential benefit of it.

Furthermore, after changing attitudes toward research participation from regarding it a burdensome and potentially dangerous to an opportunity to accessing the latest scientific innovations, enrolling in a trial could be seen as a way to receive better monitoring or medical care. This will influence the researcher's obligation to keeping in mind the pitfalls of this process, for example, distributing the burden of research on vulnerable population or underrepresentation of minorities, elder people, and those with orphan diseases [20].

In genomic researches, there is a special consideration in assessing risks and benefits of participation, especially in working with ethnic groups. Researchers who are working with minority and ethnic communities have special obligations to assure that the balance of risks and potential benefits in the study is acceptable. They can fulfill these obligations by involving the targeted group in planning the research project and educating them about the objectives and probable outcomes of it.

Independent ethical review of all clinical research protocols is also a necessary condition, first for minimizing concerns about researchers' conflicts of interest and second to ensure a high-quality research, to complete the research expeditiously, to protect research participants, and so forth. It is also important for a less emphasized reason, social accountability. If the research imposes risks to participants for the benefits of the society, it is the society's obligation to ensure that the people who enroll in trials, especially novel interventions such as OMICs, will be treated ethically [21]. To fulfill this principle first we need to develop laws and regulations for OMICs researches, and concurrently a plan to educate researchers and reviewers about the ethical challenges of OMICs researches [22].

Informed consent was frequently addressed in research on human participants. Historically, the heart of ethics in the research on human subjects was laid in obtaining informed consent from participants to ensure their understanding of risks and benefits of the research, the right to withdraw at any time from a study without consequences for medical care, and voluntary participation free of undue constraints. However, in OMICs research there are some inherent limits to this informed consent.

Part of this limitation is because of negligible significant immediate clinical consequence of OMICs research for an individual, the risks of these researches currently remain largely unknown; and particularly for genomics research there are relative concerns about re-consent because of the possibility of using genetic data in future for different purposes and even that a participant's genomic information can have implications for their family members, likely would not have originally consented to be part of the research study [23].

To manage these confounding issues, it is suggested to discuss with the participants about the meaning of genetic data openly, mention what information he or she would like to have, and determine any desired future interaction between the researchers and participants regarding the significant findings that could potentially be discovered [24, 25].

Respecting the research participants means that obtaining informed consent is not the final step of a research. Researchers have an ongoing obligation to respecting individuals from the time they are approached to participate, even if they refuse enrollment, throughout their participation, and even after the end of the research. This entails multiple activities and requirements such as: monitoring the health and well-being of participants, preventing or treating harms that might result from the adverse reactions or unwanted events, honoring the participant's privacy and confidentiality by activities such as de-identification of data, securing of personal information, regulation of access, and placement of firewalls, as well as interviewing participants in private spaces [26]. Furthermore, respecting means participants have permission to change their minds and to decide to withdraw without penalty.

Though confidentiality and privacy are traditionally ensured in research, the protection methods have attracted growing concern in OMICs research [27]. The assurance of confidentiality begins during the process of informed consent, but in

OMICs research this concept is not always so simple. For example, genetic information could reveal information that the person may prefer remain private from family members, employers, and insurance companies, the information may not be often clinically implicating and advancing technologies allow that a tiny amounts of genetic information be able to identify that individual or can be assumed. These new possibilities raise the important concern of the feasibility of maintaining patient privacy in genomic research [28].

Therefore, it is recommended to be as clear and truthful as possible with research participants in terms of who will have the right to access their data and for what purposes it will be used [29, 30].

As same as the right of confidentiality in OMICs research which is an essential concept, yet sometimes difficult to maintain, the right of research participants to withdraw from research is also an important and quite problematic one [31, 32]. Researchers should develop explicit strategies in this regard and make plans regarding the care of participants after withdrawing or when the trial is over [33]. In some cases, this may simply means referring the participants to a care provider; however, in some other cases, especially in new interventions such as OMICs, this may require to find creative strategies for providing access to treatments benefiting the participants.

Respect also requires providing any new information about the impact of the intervention or the participant's clinical condition to them. Keeping in mind that genetic data which has no clinical relevance today may become clinically relevant tomorrow and also clinically non-actionable ones could be personally important [34].

2.3 Social and Policy Issues

Our experiences with other health technologies than OMICs have taught us some important lessons. When using innovative interventions, there are some important factors other than scientific and technical factors which should be considered. As biotechnologies have both intended and unintended effects on science, medicine, and society, if these effects were not carefully considered and monitored, their sustainable and equitable development could be impended [35]. Also depend on identifying and addressing the social and policy issues. Several national and international reports aimed to address such factors [36, 37] and are applicable to both developed and developing countries [38]. Usually these reports framed the social and policy issues around categories such as resources, equity or distributive justice, and supervision or control.

Resource may involve the economic impacts of OMICs on different aspects of healthcare delivery, such as the process of drug discovery, development and marketing in Pharmacogenomics, and the effect of nutrigenomics on food marketing and diet programs [39].

These effects can result in increased or decreased drug and supplement developmental costs and prices; however, calculating their net effect would be hard.

Equity ultimately relates to the decisions and consequences associated with personalized health interventions [40]. In OMICs context, disease diagnosis and medical treatments may significantly improve for some subpopulations, but this may not for some others. For instance, inequity in the field of pharmacogenetics may affect drug development for certain groups who represent a small market for the pharmaceutical industry or a large but poor population [41].

This represents a point that usually may not be discussed in-depth, the decisions that are not made. In other words, this means that policy makers are responsible not only for the decisions they made but also for the decisions that are not made (the former shows ethics of commission and the latter shows ethics of omission). Therefore, as the ethics of omission tends to attract less attention, inequity may result, for example, a drug would not be developed for a small population that represents an inadequate financial incentive.

Furthermore, the existing scientific gap between the developed and developing countries may potentially be exacerbated with the introduction of OMICs, which is a representative of global inequity.

Supervision and control is necessary to address some critical questions such as: Who should decide whether a group of patients should take OMICs tests? What happens if a patient does not want to undergo testing, but still wishes to access the relevant personalized medicines?

Moreover, the availability of direct-to-consumer personal genomics tests calls for an appropriate regulatory oversight. Just a simple extension of previous risk assessment mechanisms seems not to be successful in this regard and in fact might lead to underprotection of patients and/or misdirected precaution [42]. There is therefore a pressing need to rethink the outdated models of regulation in the post genomics era [43]. Furthermore, as the boundaries between experts and lay persons as well as private and public sectors were blurred in OMICs era, a political and social lens is needed to discern the motives and values which drive the research and practice of the field. This will work in parallel and accompanied by normative bioethics.

In real life, these considerations will overlap considerably in applications of OMICs technologies and try to acknowledge the scientific uncertainties of OMICs research or putting it into the broader context of medical research [44]. However, there is a need for further discussion of the actual contexts and the ethical and policy issues in this regard.

3 Final Words

Considering the wave of OMICs innovations, their complexities and opportunities, and continuous grow, this chapter aimed to highlight why we need to critically examine our views not only in OMICs but also in bioethics. Regulatory oversight mechanisms and bioethics frameworks from the before OMICs era are not well equipped to address the complexities and nuances of OMICs medicine. It seems that there is a need for the integration of different disciplines and approaches such as social sciences, philosophical approaches to bioethics, and also OMICs scientists to actively take part in the process of ethical reasoning by identifying the fault lines in scientific practice which have bioethics significance, thus offering a sound framework for the ethics of OMICs practice.

Neglecting the pressing need to focus on the OMICs practice from the lens of bioethical analyses and reasoning within scientific, technical, social, and political contexts strips the practice of OMICs from the sound interaction between the medicine and society and also scatters the bioethics from lived realities of science and medicine, further nor does it leave room for effective policy interventions. Therefore as other thinkers noted, it is time to reexamine the old and new social, ethical, and policy issues with due attention to the decision-making processes and methodologies used to arrive at conclusions in both OMICs and bioethics [45, 46].

Considering the close interaction between bioethics and OMICs and the ways in which bioethical principles have been subject to reinterpretation in the light of the new challenges, this approach would be crucial to design effective and appropriately targeted policy interventions. This will lead to more unified strategy for enhancing ethical standard of OMICs and could be applied to all aspects of it from bench to bed. This is the point to start the debate.

References

1. Janitz M. Next-generation genome sequencing: towards personalized medicine. Hoboken: Wiley; 2011.
2. Pellegrino ED. Percival's medical ethics: the moral philosophy of an 18th-century English gentleman. Arch Intern Med. 1986;146(11):2265–9.
3. Sokol DK. Medical ethics; or a code of institutes and precepts adapted to the professional conduct of physicians and surgeons. Prov Med Surg J. 1849;13(26):711–2.
4. Rodriguez L, Galloway E. Bringing genomics to medicine: ethical, policy, and social considerations. In: Ginsburg GS, Willard HF, editors. Genomic and precision medicine: foundation, translation and implementation. 3rd ed; 2017. p. 283–97.
5. Beauchamp TL, Childress JF. Principles of biomedical ethics. New York: Oxford University Press; 1979.
6. Holm S. In: Beauchamp TL, Childress JF, editors. Principles of biomedical ethics. 5th ed: Oxford University Press; 2001. p. 454. isbn:0-19-514332-9.
7. Beauchamp TL, Childress JF. Principles of biomedical ethics. Oxford: Oxford University Press; 2012.
8. Beauchamp TL, Childress JF. Principles of biomedical ethics. Oxford: Oxford University Press; 2009.
9. Vijverberg S, Pieters T, Cornel M. Ethical and social issues in pharmacogenomics testing. Curr Pharm Des. 2010;16(2):245–52.
10. Blasimme A, Vayena E. Becoming partners, retaining autonomy: ethical considerations on the development of precision medicine. BMC Med Ethics. 2016;17(1):67.
11. Diamandis EP, Li M. The side effects of translational omics: overtesting, overdiagnosis, overtreatment. Clin Chem Lab Med. 2016;54(3):389–96.
12. Bagheri A, Afshar L. Research: international collaboration. In: Ten Have H, editor. Encyclopedia of global bioethics. Dordrecht: Springer; 2015. p. 1–11.

13. de Nuremberg C. Reprinted from trials of war criminals before the Nuremberg military tribunals under control council law no. 10, vol. 2. Washington, DC: US Government Printing Office; 1949. p. 181–2.
14. World Medical Association. World medical association declaration of Helsinki. Ethical principles for medical research involving human subjects. Bull World Health Organ. 2001; 79(4):373–4.
15. World Medical Association. World medical association declaration of Helsinki: ethical principles for medical research involving human subjects. JAMA. 2013;310(20):2191–4.
16. Emanuel EJ, Wendler D, Grady C. An ethical framework for biomedical research. In: Emanuel EJ, editor. The Oxford textbook of clinical research ethics. Oxford: Oxford University Press; 2008. p. 123–35.
17. Emanuel EJ, et al. What makes clinical research in developing countries ethical? The benchmarks of ethical research. J Infect Dis. 2004;189(5):930–7.
18. El Setouhy M, et al. Moral standards for research in developing countries from "reasonable availability" to "fair benefits". Hast Cent Rep. 2004;34(3):17–27.
19. Vanderpool HY. The ethics of research involving human subjects: facing the 21st century. Frederick: University Publishing Group; 1996.
20. Wilfond B, Ross LF. From genetics to genomics: ethics, policy, and parental decision-making. J Pediatr Psychol. 2008;34(6):639–47.
21. Samuel GN, Farsides B. Public trust and 'ethics review' as a commodity: the case of Genomics England Limited and the UK's 100,000 genomes project. Med Health Care Philos. 2018;21(2):159–68.
22. Bergmann MM, et al. Bioethics in human nutrigenomics research: European Nutrigenomics Organisation workshop report. Br J Nutr. 2006;95(5):1024–7.
23. Caulfield T, et al. Research ethics recommendations for whole-genome research: consensus statement. PLoS Biol. 2008;6(3):e73.
24. Ayuso C, et al. Informed consent for whole-genome sequencing studies in the clinical setting. Proposed recommendations on essential content and process. Eur J Hum Genet. 2013; 21(10):1054–9.
25. Budin-Ljøsne I, et al. Dynamic consent: a potential solution to some of the challenges of modern biomedical research. BMC Med Ethics. 2017;18(1):4.
26. Bijlsma RM, et al. Managing unsolicited findings in genomics: a qualitative interview study with cancer patients. Psycho-Oncology. 2018;27(4):1327–33.
27. Kaye J. The tension between data sharing and the protection of privacy in genomics research. Annu Rev Genomics Hum Genet. 2012;13:415–31.
28. Kaye J, et al. Ethical implications of the use of whole genome methods in medical research. Eur J Hum Genet. 2010;18(4):398–403.
29. Lin Z, Owen AB, Altman RB. Genetics. Genomic research and human subject privacy. Science. 2004;305(5681):183.
30. Rahimzadeh V, et al. Key implications of data sharing in pediatric genomics. JAMA Pediatr. 2018;172(5):476–81.
31. Holm S. Withdrawing from research: a rethink in the context of research biobanks. Health Care Anal. 2011;19(3):269–81.
32. Boddington P. Ethical challenges in genomics research. Berlin: Springer; 2012.
33. Williams JK, Anderson CM. Omics research ethics considerations. Nurs Outlook. 2018;66(4): 386–93.
34. Hsu AR, et al. A pragmatic consideration of ethical issues relating to personal genomics. Am J Bioeth. 2009;9(6–7):1–2.
35. Hewitt JE. The ethical, legal and regulatory issues associated with pharmacogenomics: systematically quantifying the literature. J Law Med. 2018;25(3):782–93.
36. World Health Organization. Genomics and world health: report of the advisory committee on health research. Geneva: WHO; 2002.
37. Lipton P. Pharmacogenetics: the ethical issues. Pharmacogenomics J. 2003;3:14.

38. Pang T. Pharmacogenomics and personalized medicine for the developing world - too soon or just-in-time? A personal view from the World Health Organization. Curr Pharmacogenom Personal Med. 2009;7:149–57.
39. Bragazzi NL. Situating nutri-ethics at the junction of nutrigenomics and nutriproteomics in postgenomics medicine. Curr Pharmacogenom Personal Med. 2013;11(2):162–6.
40. Ozdemir V, Someya T. Editorial [a Transdisciplinary forum for study of individual and population variability in response to health interventions and personalized medicine]. Curr Pharmacogenom Personal Med. 2009;7:146–8.
41. Lise L, et al. Integrating anticipated Nutrigenomics bioscience applications with ethical aspects. OMICS. 2008;12(1):1–16.
42. Prainsack B, et al. Personal genomes: misdirected precaution. Nature. 2008;456(7218):34–5.
43. Ozdemir V, et al. Risk assessment and communication tools for genotype associations with multifactorial phenotypes: the concept of "edge effect" and cultivating an ethical bridge between omics innovations and society. OMICS. 2009;13(1):43–61.
44. Corrigan OP. Pharmacogenetics, ethical issues: review of the Nuffield Council on Bioethics report. J Med Ethics. 2005;31(3):144–8.
45. Ozdemir V, Knoppers BM. One size does not fit all: toward "upstream ethics"? Am J Bioeth. 2010;10(6):42–4.
46. Kato K, Kano K, Shirai T. Science communication: significance for genome-based personalized medicineûa view from the Asia-Pacific. Curr Pharm. 2010;8:93–6.

Index

B. Arjmand (ed.), *Genomics, Proteomics, and Metabolomics*, Stem Cell Biology and Regenerative Medicine, https://doi.org/10.1007/978-3-030-27727-7

FSC
www.fsc.org
MIX
Papier | Fördert
gute Waldnutzung
FSC® C083411

Kolibri 360 GmbH
Ferdinand-Jühlke-Straße 7
99095 Erfurt, Deutschland
produktsicherheit@kolibri360.de